NEW INSIGHTS ON GAMMA RAYS

Edited by **Ahmed M. Maghraby**

New Insights on Gamma Rays
http://dx.doi.org/10.5772/62969
Edited by Ahmed M. Maghraby

Contributors
Özge Çelik, Salih Mustafa Karabidak, Marcia Nalesso Costa Harder, Valter Arthur, Allen Torbert, Galina Yakubova, Aleksandr Kavetskiy, Steve Prior, Carlos Navia, Markus Raymond Zehringer, Li Zhang, Houdun Zeng, Matthieu Hamel, Frédérick Carrel, Aura María Blandón Osorio, Clináscia Rodrigues Rocha Araújo, Geone Maia Correa, Viviane Gomes Costa Abreu, Thiago de Melo Silva, Patrícia Machado de Oliveira, Antônio Flávio de Carvalho Alcântara

Published by InTech
Janeza Trdine 9, 51000 Rijeka, Croatia

Publishing Process Manager Iva Simcic
Technical Editor SPi Global
Cover InTech Design team

First published May, 2017
Printed in Croatia
Legal deposit, Croatia: National and University Library in Zagreb

Additional hard copies can be obtained from orders@intechopen.com

New Insights on Gamma Rays, Edited by Ahmed M. Maghraby
p. cm.
Print ISBN 978-953-51-3161-8
Online ISBN 978-953-51-3162-5

For EU product safety concerns:
IN TECH d.o.o., Prolaz Marije Krucifikse Kozulić 3, 51000 Rijeka, Croatia,
info@intechopen.com or visit our website at intechopen.com.

Meet the editor

Ahmed M. Maghraby graduated from Cairo University, Faculty of Science, Egypt, in 1995 and started his career as a scientist at National Institute of Standards (NIS), Egypt. He received his MSc degree in 1999 and his PhD degree from Cairo University in 2003 for the study on the calibration and cauterization of organic material for use in radiation dosimetry using electron paramagnetic resonance (EPR). In 1999, he worked as a guest researcher in the National Institute of Standards and Technology (NIST) in Maryland, USA, and as a visiting scientist in Dartmouth College (EPR center of viable system) in New Hampshire, USA, in 2008. Also he has worked as a guest researcher in The Physikalisch-Technische Bundesanstalt (PTB), Germany's national metrology institute, in 2006. He has led and participated in several international research projects on radiation measurements. His research interest is the ionizing radiation metrology, especially ionization chambers, EPR, and thermoluminescence dosimetry (TLD). He is a reviewer for several peer-reviewed journals and has supervised a number of master and PhD theses.

Contents

Preface

Gamma radiation, the unseen energy that affected human life over decades, was discovered by the French chemist and physicist Paul Villard in 1900, 4 years after the discovery of the natural radioactivity by Henry Becquerel. X-rays were discovered by Wilhelm Roentgen in 1995; shortly, two types of radiation were discovered before the discovery of gamma radiation, alpha particles by Ernest Rutherford in 1899 and beta particles by Henry Becquerel in 1896, and hence the third type of radiation was named by Rutherford as "gamma rays" in 1903 by analogy with alpha and beta rays.

Gamma rays are highly energetic electromagnetic radiation and like other things have many useful applications and some hazards if improperly used. A day after another, new applications of gamma rays emerge, leading to more convenience for the human life and help in facing challenges in different disciplines such as in health, economy, culture, environment, and many others. Hence, there is always a need to know more about gamma rays.

This book furnishes information regarding some applications of gamma rays. It contains three sections: the first section "Gamma Ray Spectrometry" describes the use of gamma ray spectrometry in the investigation of environmental and food samples, followed by a review of studies on the dead time in gamma ray spectrometry with a newly suggested model for the evaluation of the dead time, and finally a chapter on plastic scintillators and pseudo-gamma spectrometry.

The second section, which is titled "Gamma Rays in Space," is composed of two chapters: the first gives an overview of gamma rays from space highlighting the gamma ray bursts, and the second one is concerned with the extragalactic gamma ray background and the theories describing its origin.

The last section entitled "Recent Applications of Gamma Rays" is composed of six subsections on different and sometimes unusual applications of gamma radiation such as mutation breeding, analysis of soil and minimizing the problem of allergen protein in food in addition to a detailed review of the effect of gamma rays on essential oils.

The editor thanks all authors of this book for their valuable contributions and for their patience throughout the editing processes and in the same time wishes that readers of this book will find it useful and interesting. Finally, I hope this book adds a value to the scientific library.

Prof. Dr. Ahmed M. Maghraby
Ionizing Radiation Metrology Laboratory
National Institute of Standards, Giza, Egypt

Section 1

Gamma Rays Spectrometry

Chapter 1

Gamma-Ray Spectrometry and the Investigation of Environmental and Food Samples

Markus R. Zehringer

Additional information is available at the end of the chapter

http://dx.doi.org/10.5772/67099

Abstract

Gamma radiation consists of high-energy photons and penetrates matter. This is an advantage for the detection of gamma rays, as gamma spectrometry does not need the elimination of the matrix. The disadvantage is the need of shielding to protect against this radiation. Gamma rays are everywhere: in the atmosphere; gamma nuclides are produced by radiation of the sun; in the Earth, the primordial radioactive nuclides thorium and uranium are sources for gamma and other radiation. The technical enrichment and use of radioisotopes led to the unscrupulously use of radioactive material and to the Cold War, with over 900 bomb tests from 1945 to 1990, combined with global fallout over the northern hemisphere. The friendly use of radiation in medicine and for the production of energy at nuclear power plants (NPPs) has caused further expositions with ionising radiation. This chapter describes in a practical manner the instrumentation for the detection of gamma radiation and some results of the use of these techniques in environmental and food investigations.

Keywords: gamma-ray spectrometry, neutron activation analysis, radioactive contamination, radiocaesium, radiostrontium

1. Gamma spectrometric equipment for the control of food and environmental samples

1.1. Theory of gamma spectrometry

Gamma rays are electromagnetic radiation and are part of photon radiation. They are produced when transitions between excited nuclear levels of a nucleus occur. Delayed gamma rays are emitted during the decay of the parent nucleus and often follow a Beta decay. There can be many transitions between energy levels of a nucleus, resulting in many gamma-ray lines. The typical wavelength is 10^{-7} to 10^{-13} m, corresponding to an energy range of 0.01–10 MeV.

Gamma rays can be detected through their interaction with matter. There are three main processes: photoelectric absorption, Compton scattering and pair production. The photoelectric effect occurs when a gamma ray interacts with an electron of an inner shell of an atom and a photoelectron is emitted. This is the most important effect for the detection of gamma rays with semiconductor detectors. The effect of Compton scattering describes the interaction of a gamma ray with matter when some of its energy is transferred to the recoil electron. The energy transmitted is a function of the scattering angle. Therefore, the Compton effect results in a broad range of gamma-ray energies, which gives a continuous background in the gamma spectrum. Pair production is the third effect when a gamma ray is absorbed by matter and loses energy to produce an electron/positron pair. This effect only occurs when gamma rays have more than 1.02 MeV energy, twice the rest mass energy of an electron (0.551 MeV) [1–4].

1.2. Semiconductor detectors

Until the mid-1970s, no germanium could be produced of the desired purity. The purity required for large-volume detectors could only be produced by doping germanium crystals with n-type impurities, such as lithium (Ge(Li)-detectors). Later on, pure germanium crystals became available in n-type or p-type form and of closed-end coaxial or planar geometry and as bore-hole crystals. n-type detectors cover an energy range from about 10 keV to 3 MeV, while p-type detectors range from 40 keV to 3 MeV. p-type detectors with a carbon fibre or beryllium window instead of the aluminium end cap are best to detect energies below 100 keV.

1.3. Requirements for proper gamma spectrometry

The minimum detectable activity (MDA) of a Ge-detector depends on its energy resolution, the efficiency of the crystal, peak/Compton-factor, background, measuring time, sample geometry, self-absorption and the emission probabilities of the gamma emission lines of the radionuclide. Information can be obtained from the homepages of the providers, such as Ortec, Canberra and others [5–7].

1.3.1. Detector calibration

Ge-detectors are calibrated for the energy response of the multi-channel analyser, peak resolution and counting efficiency. Normally, gamma spectroscopists use commercially available calibration sources containing a mix of gamma-nuclides, which cover the whole energy range. Such nuclides are ^{210}Pb or ^{241}Am, ^{109}Cd, ^{139}Ce, ^{57}Co and ^{60}Co, ^{134}Cs or ^{137}Cs, ^{88}Y, ^{85}Sr. Such mixes cover an energy range from 46 keV (^{210}Pb) to 1836 keV (^{88}Y). The disadvantage is that some of the nuclides have short half lives (e.g. ^{85}Sr has a half-life of 65 days) and therefore such calibration mixes only can be used for a year. Sometimes it might be better to use a mix of a low-energy nuclide (e.g. ^{241}Am) and ^{152}Eu, which disintegrates slowly (half-life of 13.5 years) and shows a multiplicity of emission lines from 122 to 1528 keV. The disadvantage is the summing effects of ^{152}Eu, which require correction (e.g. using software based on Monte Carlo simulations). Furthermore, the calibration of peak resolution and efficiency of the sample geometry is necessary; this is performed with the same calibration sources. We use calibration sources with ^{241}Am/^{152}Eu of different sample geometries. They are solidified by gelation and have a

http://dx.doi.org/10.5772/67099

density of 1.0 g/mL. Such sources are available, e.g. at Czech Metrology Institute at Prague [7]. Peak resolution should be tested on a regular basis together with the energy calibration. The peak shape is close to a Poisson distribution. For more counts, the distribution is closer to a Gaussian shape. The peak resolution is given by the quotient of FWTM (full width at tenth maximum) versus FWHM (full width at half maximum). Ge-detectors show resolutions of typically 1–2.5 keV. The software for the recording and analysis of pulse high spectra is available from Canberra, Ortec products, Oxford instruments, etc. Interwinner software from ITEC is a commonly used software in Germany and Switzerland [8].

1.3.2. *Background*

It is absolutely necessary to know the background of the Ge-detector system. This depends on the shielding of the detector and the background of the laboratory. We use shielding with 10 cm of lead and an inner layer of copper 5-mm thick. Before our laboratory was built, radiation-poor materials for the construction of the walls, soil and ceiling were sought. We analysed different components, such as gravel, sand, white cement and additives from different producers. Our choice for gravel and sand was a local producer in the Swiss Alps. The cement was from Dyckerhoff in Denmark. With this effort, we could reduce the background of our counting laboratory from 70 to 20 nSv/h. Nevertheless, background is still present. Incoming cosmic muons are not suppressed. They can be reduced by building an anticoincidence chamber over the detector. Vojtyla et al. could reduce the background by a factor of 2.2 [9]. Another approach is described by Seo et al., using Marinelli beakers of aluminium and purging the surrounding air of the detector reduced the background [10]. The background has to be measured periodically for each geometry to this end; a sample container with deionised water is placed on the detector and counted over a weekend. The background depends on the counting geometry and the matrix.

1.3.3. *Attenuation effects*

Photons may be absorbed by the sample matrix and therefore do not reach the detector. This effect depends on the elemental composition of the sample and its density. The photon attenuation effect is not negligible for photons with lower energies and for high-volume sample geometries. It has to be taken into account with the Gamma spectrometry software or software based on Monte Carlo simulations.

1.3.4. *Coincidence summing effect*

This effect occurs for all radionuclides emitting at least two photons in sequence and is a function of the source-detector distance and the detector efficiency. With a 50% Ge-detector, more coincidence summing is recorded than with a 20% detector. To avoid this effect, the samples may be counted a certain distance away from the detector.

1.3.5. *Dead time*

Samples of high activities may lead to a loss of peak counts. This effect occurs when the pulse processing electronics are slower than the frequency of the incoming photons. This leads to a dead time of the detector. Normally, environmental and food samples do not show such high activities that result in dead times of the detector. Dead time can be avoided by counting the sample at a well defined distance from the detector. Such constellations can be important when the samples have to be analysed for an emergency case.

1.4. Best sample geometries for gamma-ray analyses

MDA is a function of the detector efficiency and the sample weight. The best gamma-ray efficiencies are achieved with Marinelli beakers of 1 or 2 L, as the gamma rays of the sample interfere on top and on the sides with the Ge-crystal, gaining more efficiency. This geometry is best when large sample amounts of water, soil, vegetation, food, etc. are available. Marinelli geometries have to be calibrated carefully and coincidence summing has to be corrected. For small sample amounts, dishes with volumes of 32 and 77 mL (12 or 24 mm height and 6.5 cm in diameter) might be used. In small sample devices, the attenuation of gamma rays by the sample matrix is remarkably decreased. The disadvantage is the small sample load of 30–80 g. Other geometries commonly used are beakers of 250 and 500 mL volume. To enhance the sensitivity of the gamma-ray spectrometry, samples containing water may be freeze-dried. For milk samples, a concentration factor of eight can be achieved by freeze-drying. Soil, vegetation and food samples should also be dried (e.g. at 120°C). Soil samples are ground and sieved to eliminate large particles, such as stones or root parts. Further practical advice is given in other sources [11, 12].

1.5. Interpretation of gamma spectrometry data of natural radionuclides

Dose-relevant radionuclides of the natural decay series of ^{238}U, ^{232}Th and ^{235}U are nuclides from uranium, radium, thorium, actinium, lead and polonium. Relevant criteria are the half-life and the dose coefficients of these radionuclides. With a few exceptions, these radionuclides can be detected via the gamma emissions of their daughter nuclides. The exceptions are ^{7}Be, ^{40}K, ^{223}Ra, ^{210}Pb, ^{231}Pa. The detection of ^{226}Ra and ^{224}Ra needs a secular equilibrium between the mother nuclide and its daughters. This can be reached when the sample is packed gas tight for at least the sevenfold half-life of the corresponding radon nuclide prior to the gamma analysis. This equates to 20 days in the case of ^{226}Ra, or 7 min for ^{224}Ra.

$$\begin{array}{l} ^{226}Ra \xrightarrow{-\alpha} {}^{222}Rn \xrightarrow{-\alpha} {}^{218}Po \xrightarrow{-\alpha} {}^{214}Pb \xrightarrow{-\alpha} {}^{214}Bi \xrightarrow{-\beta} {}^{214}Po \\ ^{224}Ra \xrightarrow{-\alpha} {}^{220}Rn \xrightarrow{-\alpha} {}^{216}Po \xrightarrow{-\alpha} {}^{212}Pb \xrightarrow{-\beta} {}^{212}Bi \xrightarrow{-\beta} {}^{208}Tl \xrightarrow{-\beta} {}^{208}Pb \end{array} \tag{1}$$

After reaching secular equilibrium, the activities of the daughters of ^{222}Rn and ^{226}Ra can be set equal to the activities of ^{214}Pb and ^{214}Bi. Often, the direct determination of ^{226}Ra is not possible due to the major interference with the gamma line of ^{235}U around 186 keV. Therefore, this is the best method to detect radium using gamma-ray spectrometry. The same applies for the system ^{228}Th/^{224}Ra and their daughters ^{212}Pb and ^{212}Bi. Here, equilibrium is reached within minutes due to the very short half-life of ^{220}Rn.

Other mother-daughter systems can be used for the determination of ^{232}Th, ^{228}Ra (a pure β-emitter), ^{227}Ac and ^{238}U.

http://dx.doi.org/10.5772/67099

$$^{232}Th \rightarrow {}^{228}Ra + \alpha \rightarrow {}^{228}Ac + \beta, {}^{227}Ac \rightarrow {}^{227}Th + \beta, {}^{238}U \rightarrow {}^{234}Th + \alpha \rightarrow {}^{234m}Pa + \beta. \quad (2)$$

Due to the very short half-lives of the daughters, these radionuclides are already at equilibrium. **Table 1** shows the adequate choice of gamma emission lines for relevant natural radionuclides. In natural uranium, the activity ratio of $^{238}U/^{235}U$ is 21.7.

Radionuclide		Energy (keV)	Emission probability ε (%)	Interferences
^{7}Be	Direct	477.61	10.3	
^{40}K	Direct	1460.8	10.67	
^{226}Ra	Direct	186.2	3.5	^{235}U (185.72 keV; 57.2%)
	^{214}Pb	295.21	18.2	^{211}Bi (351.06 keV; 12.91%)
	^{214}Pb	351.92	35.8	
	^{214}Bi	609.32	44.6	
	^{214}Bi	1120.3	14.8	
	^{214}Bi	1764.5	15.4	
^{228}Ra	^{228}Ac	338.32	11.3	
	^{228}Ac	911.21	26.6	
	^{228}Ac	968.97	15.8	
^{228}Th	^{224}Ra	240.99	4.0	^{214}Pb (241.98; 7.12%)
	^{212}Pb	238.63	43.3	^{228}Ac (583.41 keV; 0.114%)
	^{212}Pb	300.09	3.3	
	^{208}Tl	277.36	2.3	
	^{208}Tl	583.17	30.5	
	^{208}Tl	860.56	4.5	
^{227}Ac	^{227}Th	235.97	12.1, 11.2	
	^{227}Th	256.5	7.0	
	^{223}Ra	269.5	13.7	
^{235}U	Direct	143.76	10.96	^{226}Ra (186.1; 3.51%)
		163.33	5.08	^{228}Ac (204.10; 0.171%)
		185.72	57.2	
		205.31	5.01	
^{238}U	^{234}Th	63.28	4.3	^{232}Th (63.81; 0.267%)
	^{234}Th	92.37	2.5	Weak line
	^{234}Th	92.79	2.4	Weak line
	^{234M}Pa	766.37	0.21	
	^{234M}Pa	1001.03	0.84	
^{210}Pb	Direct	46.54	4.2	Weak line
^{223}Ra	Direct	154.21	5.6	
	^{211}Bi	269.46	13.7	
	^{219}Rn	351.07	12.9	
	^{219}Rn	271.23	10.5	
		401.81	6.5	
^{231}Pa	Direct	300.07	2.5	
		302.67	2.2	

Table 1. Common used emission lines for the detection of natural radionuclides with gamma-ray spectrometry. Emission probabilities are mean values from different sources [13–17].

When analysing for natural radionuclides, it is very important to know and to reconsider the background of the gamma system. Prominent radionuclides in the background are radon and its daughter nuclides from underground radiation. The background has to be determined for each geometry and has to be subtracted from the sample. We determine the background radiation using gamma spectrometry with sample containers containing deionised water, to take into account that the sample matrix also absorbs a part of the background radiation.

2. Instrumental neutron activation analysis

2.1. Principle

Instrumental neutron activation analysis (INAA) is based on the production of short-lived radionuclides by nuclear reactions. Most frequently, reactor neutrons (i.e. thermal neutrons) are used to activate many nuclides to produce radioactive nuclides. The efficiency of the irradiation process depends on the flux density of the neutrons and the cross section of the nuclear reaction of the irradiated nucleus. Typically, the thermal neutrons required for INAA are generated in a nuclear reactor [18, 19]. For our experiments, we used the reactor at the University of Basel (AGN-211-P), which is a light water-moderated swimming pool reactor. The compact core contained 2.2 kg of highly enriched uranium and a graphite reflector around the core. This uranium gave a thermal neutron flux of 3.8×10^{10} n/cm^2/s at a power of 2 kW. The insertion of samples into the core was possible over a cannula (brown cylinder above the reactor in **Figure 1**) through the so-called glory hole. The glory hole consists of an air-filled pipe of a diameter of 1″ (2.5 cm), which goes from above the pool down and through the centre of the reactor.

Figure 1. Equipment for NAA. Swimming pool reactor (left) gamma-ray spectrometer with lead shielding (middle) gamma-ray spectrum (above right) curry sample and sample device for the irradiation [4].

http://dx.doi.org/10.5772/67099

2.2. Operational procedure

The detection limit of gamma spectrometry is given by the half-life of the activated nuclides and the underlying Compton background from highly activated nuclides, such as sodium or chloride. Gold foils are used as the internal standard for each sample and are set on top of each sample. A sample series of 12 samples each of 1–2 g material was irradiated for 30 min at a power of 2 kW. Each irradiation place in the neutron field of the reactor was calibrated with coagulated salt solutions containing a known amount of the analyte and a corresponding gold foil. The response factors of each analyte to its gold foil for each place in the neutron field were then calculated (comparator method).

2.3. Common applications of INAA

2.3.1. Determination of total bromine content in food samples

The total bromine content of food, such as tea, coffee, dried mushrooms, vegetables and spices, gives information about the use of methyl bromide, a fumigant. The application of methyl bromide results in residues of bromide. This bromide can be activated to the gamma-active compound ^{82}Br (half-life of 35 h) by neutrons and analysed with gamma spectrometry. After irradiation of 30 min, the samples have to be cooled down for several hours (for the disintegration of activated sodium and chloride nuclides). The gamma analysis takes 15 min.

$$Activation\ process: {}^{81}Br + n \rightarrow {}^{82}Br \qquad Decay\ process: {}^{82}Br \rightarrow {}^{82}Kr + \beta + \gamma. \tag{3}$$

According to **Table 2**, many objections had to be executed for spices and dried mushrooms, which were treated with methyl bromide. Our last investigation of tea resulted in one objection. Since some years, the use of methyl bromide as a fumigant has been rare. Other fumigants, such as sulfuryl fluoride, hydrogen cyanide and phosphines, have become more important [20].

2.3.2. Determination of total iodine content in food

Algae and other food samples rich in iodine were irradiated to determinate the total content of iodine. Iodine is essential for the production of thyroid hormones and prevents goitre. In most European countries, people suffer from an iodine deficiency. The iodine level can be increased by the consumption of iodine-enriched food and dietary supplements. However, high levels of iodine (i.e. over 500 μg/kg) can result in hyperthyreosis. Therefore, the range of tolerance for iodine is narrow and it is important to declare the correct iodine content for food.

About 1 g of sample can be activated with reactor neutrons (30 min, 2 kW). The radioactive product ^{128}I is analysed directly using a gamma spectrometer

$$Activation\ process: {}^{127}I + n \rightarrow {}^{128}I \qquad Decay\ process: {}^{128}I \rightarrow {}^{128}Xe + \beta + \gamma. \tag{4}$$

The half-life of ^{128}I is only 25 min; therefore, the samples had to be counted immediately after the activation. This is unfavourable regarding the background of other activated ions and results in higher detection limits. The total iodine content of fish, seafood, algae and dietary supplements can be analysed [22, 23].

Year/food	Spices	Dried mushrooms	Tea	Coffee	Chocolate	Rice
1988	10 (172)					
1989	4 (34)	7 (48)				
1990					0 (30)	
1991	5 (28)		0 (27)			
1992	2 (57)			0 (5)		
1994	0 (30)					0 (24)
1995	0 (33)					
1996		1 (33)				
1998		0 (15)				
2001	1 (26)					
2002			2 (33)			
2006			0 (17)			
2009			1 (40)			

Objections, which prove a use of the fumigant methyl bromide, show a bromine content over the limit. Limit values are 50 mg/kg (coffee, tea) and 100 mg/kg (spices mushrooms) according to the Swiss Ordinance on contaminants and constituents in Food [21]. The number of investigated samples is given in brackets.

Table 2. INAA analyses of food samples for total bromine content.

2.3.3. Determination of flame-retarding agents in plastics

INAA can be used as a screening analysis for flame-retarding agents in plastic materials, such as decabromo-bis-phenylether or tetrabromo-bisphenol A. The activation and decay process are the same as for the bromine analysis in food samples. The INAA gives information about the total content of brominated flame-retarding agents. We used INAA as a screening analysis and samples containing a high amount of bromine were detected and then analysed with gas chromatography to determine the amount of different flame-retarding compounds [24–26].

2.3.4. Determination of U and Th in suspended matter, sediment and soil samples

About 1 g samples of dried and ground material (e.g. freeze-dried suspended matter) were irradiated for 30 min at a power of 2 kW. After a cooling period of 2 h, the samples were analysed with a gamleast 30 min [27]

$$\begin{aligned} &\textit{Activation processes}: {}^{238}U + n \rightarrow {}^{239}U + \beta + \gamma \quad \textit{and} \quad {}^{232}Th + n \rightarrow {}^{233}Th \\ &\textit{Decay processes}: {}^{239}U \rightarrow {}^{239}Np + \beta + \gamma \rightarrow {}^{239}Pu + \beta + \gamma \\ &\textit{and } {}^{233}\mathrm{Th} \rightarrow {}^{233}\mathrm{Pa} + \beta + \gamma \rightarrow {}^{233}\mathrm{U} + \beta + \gamma \end{aligned} \tag{5}$$

http://dx.doi.org/10.5772/67099

3. Gamma-ray sources in the environment

Materials that emit radioactive rays are called radioactive sources. We distinguish between naturally occurring radioactive material (NORM) and technologically enriched naturally occurring radioactive material (TENORM) on the one hand and artificially produced radioactive sources on the other. Radionuclides may emit different rays, such as alpha, beta and gamma rays. Most α- and β-decays are accompanied by γ-rays. There are only a few important exceptions, such as ^{210}Po, ^{63}Ni, or ^{90}Sr, which are pure α- or β-emitters. Therefore, many important β -nuclides can be detected with gamma spectrometry.

3.1. First use of natural radioactivity

Soon after the discovery of radioactivity by Henry Becquerel, Pierre and Marie Curie and others, when radium became available, the production and the commercial use of TENORM began. Radium and thorium were used as remedies to cure many diseases. Underwear and wool, soap, lipstick, hair shampoo, toothpaste, suppositories, soda drinks, butter, etc. were spiked with NORM or TENORM. The negative effects of TENORM went visible only decades later. Quacks, such as William Bailey, earned their money by dealing with radioactive sources as medicinal drugs. It was therapy with radithor (a mixture of ^{226}Ra and ^{228}Ra) that led to the tragic death of Eben McBurney Byers [28]. Another tragedy was the "radium girls" from New Jersey. Many young women became ill or died from painting watch dials with radium. These and many more cases became public and led to the decline of the popularity of radioactivity [29]. Today, radon therapy in radon water, inhalation of radon air in tunnels or drinking of radon water remain the few existing applications of NORM for health cures against chronic diseases, rheumatic diseases and Morbus Bechterew. The health effects of radon are well described, but are not fully understood [30]. Such dubious items of the past are sometimes still present in households (see Section 3.3).

3.2. Natural radioactivity in food

Some natural radionuclides from the natural decay series of uranium and thorium enter the food chain. The alpha nuclide polonium-210 (^{210}Po), a product of the decay series of uranium-238 (^{238}U), is enriched in the intestinal tract of mussels and fish. Lead-210 (^{210}Pb), radium and thorium nuclides are present in cereals. In addition, spices and salt may contain elevated levels of radium and potassium-40 (^{40}K). Generally, potassium-rich food is also rich in ^{40}K (e.g. tea, vegetables). A special case is Brazil nuts, which are enriched in radium from soil. This is well described in [31]. Tap water may contain uranium, radium and their daughter nuclides depending on the local geological situation [32].

3.3. Radioactive sources in consumer products

Remnants from the application of natural radionuclides in the past century may be present in households even today. Our laboratory maintains a collection of radioactive objects that people brought in for investigation or disposal.

The use of thorium in flame detectors is widespread: for instance, in dials with radium in watches or on dial-plates for military use, coloured glass pearls or drinking glasses containing uranium oxides, wall tiles with uranium oxides, etc (**Table 3**). The finders of such items are encouraged to bring them to a specialised laboratory or to a collecting point for radioactive materials. The included radioactive material may be harmful.

Consumer product	Radionuclide(s)	Radionuclide content range
Radio luminous timepieces	^{3}H ^{147}Pm ^{226}Ra	4–930 MBq 0.4–4 MBq 0.07–170 kBq
Marine compass	^{3}H ^{226}Ra	28 MBq 15 kBq
Aircraft luminous safety devices	^{3}H ^{147}Pm	10 kBq 300 kBq
Static eliminators	^{210}Po	1–19 MBq
Dental products	^{nat}U	up to 4 Bq
Gas mantles	^{232}Th	1–2 kBq
Welding rods	^{232}Th	0.2–1.2 kBq
Optical glasses Ophthalmic lenses [35]	^{232}Th	5–75 Bq
Glassware: vaseline glass, canary flint glass	^{nat}U	100 kBq
Lamp starters	^{85}Kr	0.6 kBq
Smoke detectors	^{241}Am	37 kBq
Electron capture detectors	^{63}Ni	370 kBq
Drinking devices "Radium Drinkkur"	^{226}Ra, (^{222}Rn)	100 MBq
Wall tiles, ceramics	$^{nat}UO_3$	50–500 kBq
Granitic surfaces	^{nat}U	5–10 kBq/kg
Cardiac pacemaker [36]	^{239}Pu	113 GBq

Table 3. Consumer products containing radioactive materials (modified after Ref. [37]).

Incandescent gas mantles are in use without the knowledge of any possible danger. They contain ^{232}Th-oxides used to produce a bright light, which may be inhaled when the gas mantle disintegrates. In Germany and Switzerland, these gas mantles have been banned from the market. Attention has to be given to imports of products from the Far East. They may still contain thorium oxides.

A special case is the "Radium Drinkkur" (radium drinking device). It was used at the start of the twentieth century. The "Drinkkur" contained a small piece of pitch blend as a radium source (e.g. 100 MBq). The idea was to enrich drinking water with radon by emanation. Radon is said to be a remedy against rheumatics. Our own experiments have shown that a 2 month application of such a drinking therapy gave a yearly dose of 34 mSv, only from the radon. Unfortunately, radium was also released when the source was immersed into the water. Therefore, an even higher dose with additionally washed-out radium could be incorporated [33].

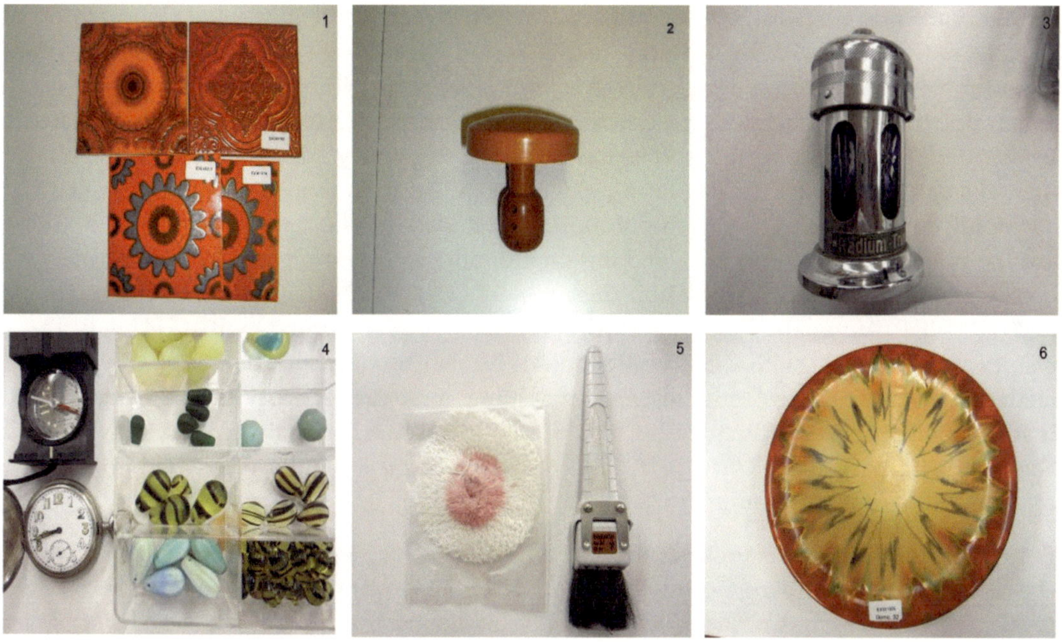

Figure 2. (1) Wall tiles; (2) pitch-blend source for radium drinking device (3); (4) watch, compass, glass pearls; (5) gas mantle and static eliminator; (6) bowl with paintings.

In the 1960s, radioactive wall tiles were discovered in Swiss households. They were produced with uranium oxide to obtain a brilliant red colour. Radiation from the walls of kitchens and toilets was of minor concern (low gamma energies), but a certain risk existed when the tiles were removed. The unavoidable dust contained uranium oxides; its inhalation had to be avoided. The Federal Office of Public Health regulated the professional drawbacks and disposal of the radioactive tiles [34] (**Figure 2**).

4. Gamma nuclides in the environment

Radioactive fallout is the main source for artificial radionuclides in the environment. In the following section, the application of gamma-ray spectrometry in Swiss environmental monitoring programs will be presented with examples [38].

4.1. Swiss monitoring programme

The Federal Office of Public Health (BAG) publishes a yearly report on the radioactivity in the environment and on radiation doses of the Swiss public. Several institutions, such as

Labor Spiez (LS), Institut de Radiophysique Appliqué (IRA), Paul Scherrer Institut of ETH Zurich (PSI), the Swiss Federal Institute of Aquatic Science and Technology (EAWAG), the National Emergency Operations Centre (NAZ), the Swiss Accident Insurance Fund (SUVA), the Swiss Federal Nuclear Safety Inspectorate (ENSI), the BAG, the European Organization for Nuclear Research (CERN), laboratories of the NPPs and some of the state laboratories of Switzerland, analyse different compartments with different techniques [39]. The main content of the reports lies in the supervision of emissions from NPPs and other industry using and producing radionuclides, the emissions from wastewater treatment plants and waste incineration plants. The report shows the results of the yearly survey of a grid of environmental sampling points, such as farms, sampling points in the vicinity of NPPs, water and air monitoring stations. Data from monitoring stations are collected and interpreted. These are the NADAM-net of the NAZ (66 automatic radiation dose meters [40]), the MADUK-net, operated by the ENSI (radiation dose meters [41]), RADAIR (air monitoring stations) and URANET (Automatic River monitoring detectors) both operated by BAG [42, 43]. These results are completed with the investigation of human tissues, such as the investigation of teeth and bones by the IRA or whole-body counting at the university hospital in Geneva and investigation of special radionuclides, e.g. ^{14}C in tree leaf samples in the vicinity of NPP's and chemical industries by the University of Berne. Based on these results, the radiation exposure of the public is estimated annually.

From the beginning, the state laboratory of Basel-City took part. The Office of Public Health chose for us three sampling points (farms) in Ticino, one farm in Basel-Country and a milk processing centre in the Canton of Jura for the yearly analysis of soil (upper 5-cm layer), grass and milk. Milk from milk distribution centres and other sites are analysed for radiostrontium as a supplement. Also, the survey of suspended matter of the River Rhine in Basel was delegated to our laboratory. Our laboratory also controls the local emissions of radioactivity. These are the wastewater of the local hospitals, the waste water treatment plant (WWTP) of Basel, ProRheno, and the incineration Plant of Basel, KVA Basel. Monitoring was started in 1993.

4.1.1. *Environmental monitoring*

In 1986, the southern parts of Switzerland were the most heavily affected by the fallout from Chernobyl. This can clearly be demonstrated by the time series of the investigated soil, grass and milk samples. There and also in elevated sampling points, such as in the Jura mountains, a remarkable increase in the radio-strontium and radio-caesium levels from global fallout was observed from 1986 to 1988. Other sampling points, such as Grangeneuve, the vicinity of the NPP's of Gösgen and Mühleberg and Leibstadt showed the normal decreasing trend of the global fallout with only slight additional fallout from Chernobyl [44]. For radiostrontium, they did not observe an increase in the contamination. Corcho et al. [45] recently published trend curves for three alpine investigated sites and two sampling sites in Swiss Mittelland. They analysed radioactive data of soil, grass and milk samples from 1994 to 2013. The effective half-lives did not depend on the altitude of the site. Radiostrontium showed quite a shorter half-life than the physical half-life. This can be explained by its migration to deeper soil layers and therefore less being available for plants. On the contrary, the effective half-life of radiocaesium is similar to its physical half-life. This is because it is fixed on clay particles in the soil and moves only slowly into the deeper soil layers [45].

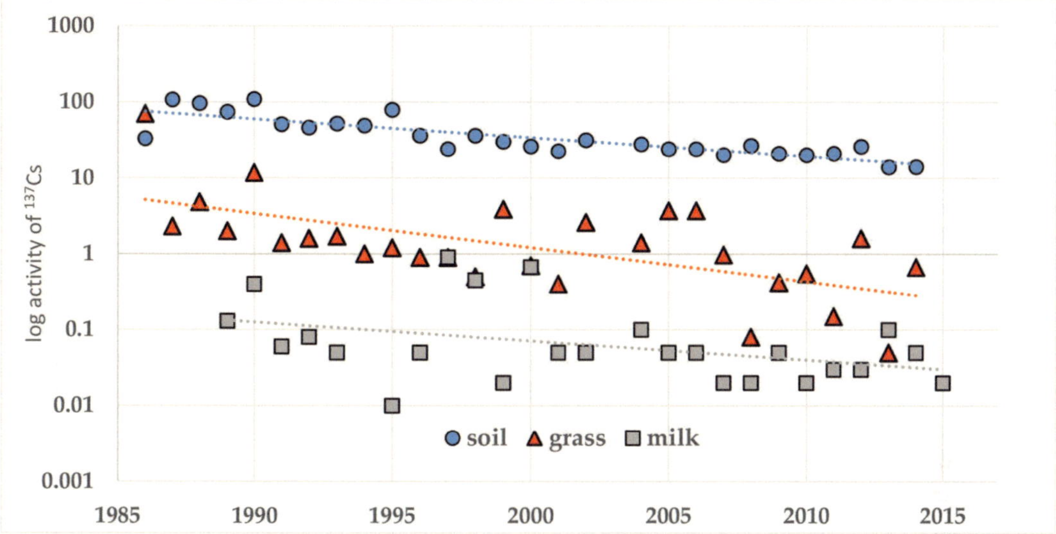

Figure 3. Activity trends of radiocaesium in soil, vegetation and cow's milk on a farm in Basel-Country. We calculated the effective half-live to 12.9 years (soil), 11.6 years (grass) and 12.0 years (cow's milk).

One of our regular sampling points is a farm in Basel-Country (450 m altitude). Data from 1986 until today are available. The data analysis of ^{137}Cs shows the following trends (**Figure 3**). The effective half-life of radiocaesium is lower than the physical one. We explain this fact by movement of the radionuclide into deeper soil layers. Therefore, the radiocaesium becomes progressively less available for the grass roots.

Such calculations are of interest when trends in the behaviour of radionuclides for food and feed are sought. They provide important information about the contamination of the food chain. The Chernobyl impact resulted in an increase in the contamination in cow's milk for about 6 years.

4.1.2. Monitoring of local emissions of radionuclides

In Basel, the main emission sources of radioactive material are the chemical/pharmaceutical industry, hospitals and some minor industries using radioactive sources. Emissions from these sources are deposited in the environment via wastewater and the air. The waste water is cleaned at the local WWTP ProRheno and two waste incineration plants, the city's incineration plant, KVA Basel and an incineration plant of the chemical industry for hazardous wastes, RSMVA of Valorec.

An important task of the state laboratory is the monitoring of the local wastewater effluents. These monitoring programmes started in 1988 and included the parameters ^{3}H (mainly used by the local chemical/pharmaceutical industries) and short-lived radionuclides used in hospitals, such as ^{131}I, ^{99m}Tc, ^{90}Y, ^{111}In, ^{177}Lu, ^{186}Re, ^{153}Sm, ^{67}Ga and others. The university hospital of Basel is specialised towards DOTATOC therapies, where mainly ^{90}Y and ^{177}Lu are used. Wastewater from the patients is collected in cool-down tanks for some weeks before being discharged to the wastewater treatment plant. Monitoring results are published yearly [46].

Tritium emissions are regularly detected in the washing water of the air filters of KVA Basel. Several times during the last 25 years, tritium was above the permitted emission limits of 6000 Bq/L and 60,000 Bq/month. However, these emissions contribute very little to the ^{3}H-level of the River Rhine. Here, the main ^{3}H sources are the NPPs and tritium producing industry in Switzerland upstream [47].

The main contamination factor of air is radiocarbon, which are used mainly by the chemical/pharmaceutical industries. In Basel, radiocarbon is emitted when waste is burned at RSMVA. Incineration takes place mainly in the night, to lower the uptake of radiocarbon in plants by photosynthesis. These emissions are controlled annually by the University of Berne. The monitoring programme of the University of Berne includes other emission sources in Switzerland, such as incineration plants, ZWILAG and NPPs [39].

4.1.3. *Behaviour of radionuclides in a WWTP*

In 2014, the influents and effluents of the WWTP ProRheno were investigated, to obtain a balance of short-lived radionuclides for medical use. The main fraction of ^{131}I was dissolved in wastewater and over 90% of the input was emitted with the treated wastewater into the River Rhine. Only a small amount was emitted via air when the sewage sludge is incinerated. For ^{177}Lu, we observed that about 60% of the incoming activity was eliminated in the WWTP, mainly by adsorption on the sewage sludge. Finally, it remained in the sludge ash, which is deposited on a nearby landfill disposal site. It disintegrates there rapidly. About 40% of the activity is emitted with the treated wastewater to the River Rhine and deposited on suspended matter and river sediment [48, 49]. The investigation of sewer sludge in the local sewage water system of Basel clearly shows that a part of the activities found at the influent of the WWTP originates from patients treated in ambulances [50].

Some of the waste from the hospitals is burnt at the incineration plant. This is proven by the activities found in the washing water of the air filters. While ^{131}I is found permanently in the low Becquerel level, other nuclides only are found sporadically. The main contamination factor at KVA Basel is ^{3}H. The sources of these emissions are not known. The Swiss Accident Insurance Fund, SUVA, supposes the source to be the accidental burning of ^{3}H-containing watches with the daily-delivered waste from households and industry.

4.1.4. *Suspended matter of the River Rhine*

Many contaminants, such as organics, metals and radionuclides, adsorb onto clay particles and are transported in a river as suspended matter. After quite a long distance or sections where the river water stands still, e.g. behind dams, the suspended particles settle onto the river sediment. Radionuclides released from NPPs are monitored by EAWAG and our laboratory at defined sampling points downstream. Suspended matter is collected either continuously by a special particle-settling chamber, or by the use of a centrifuge (i.e. non-continuous monitoring). At the river monitoring station Weil, downstream of Basel, suspended matter is collected monthly with a centrifuge. The freeze-dried and ground material is then analysed with a Ge-detector. Beside NPP-specific radionuclides, such as ^{60}Co, ^{54}Mn or ^{65}Zn, radionuclides from medicinal applications (^{131}I, ^{177}Lu and others), natural radionuclides from the decay series of U and Th and from fallout (^{137}Cs) can be detected [39, 51].

4.2. Special applications/projects

We now describe our own investigations to find representative organisms for radioactivity monitoring of the environment. Therefore, we analysed possible sample types, such as mosses, soil, grass, dust, water and wood.

4.2.1. Behaviour of radionuclides in soil filters of local drinking water production

The filtration of river water through forest soils is the most important step of drinking water production in Basel. In the context of an emergency concept for drinking water production, the question arose as to how the radioactive contaminants behave when entering the soil filter. Is there a danger of contamination of the drinking water after a nuclear accident? We analysed soil cores in one of the filtering fields for global fallout. As described in the literature, radiocaesium and plutonium remain in the upper soil layer. Radiostrontium moved deeper into the soil. For radiocaesium, we estimate two main sources: global fallout and fallout from Chernobyl. We estimated that in 1986, only about 65% of the total caesium reached the infiltration site with the infiltrated river water. The rest was bound to the suspended matter and removed by sand filtration before the infiltration step. Plutonium and perhaps radiostrontium, only originates from global fallout. We suppose that radionuclides are retained in the soil even at higher charges. For radiostrontium, the retention was lower and could have reached the groundwater. Further investigations are necessary (**Figure 4**) [52].

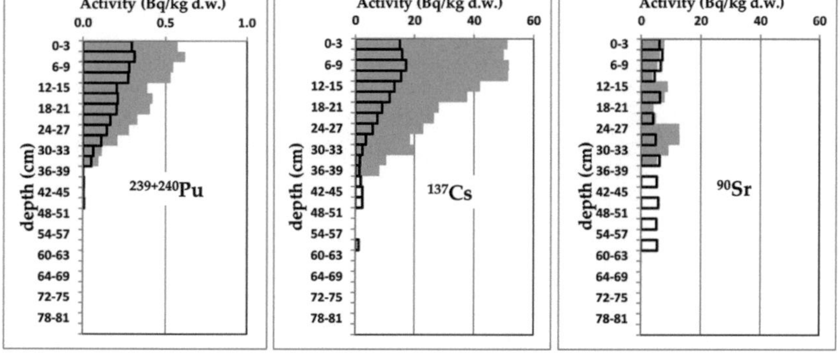

Figure 4. Soil profiles from a soil filtration site of the drinking water producer of Basel. Bars in grey are cores from the filtration site compared to a reference site outside (bars in white). From Ref. [52].

4.2.2. Fallout monitoring at Basel and vicinity

During the annual emergency exercises for an A-impact, soil samples were collected (the upper 5-cm layer) over 50 different sampling points in the city of Basel and surroundings at different altitudes (250–360 m). The analysis of over 200 soil samples with gamma-ray spectrometry resulted in an overall range of 10.6 ± 6.4 Bq/kg of ^{137}Cs.

A second project was focused on the analysis of mosses. Mosses were analysed with beta and gamma spectrometry. The analysis with gamma-ray spectrometry resulted in an overall range 2.2 ± 2.6 Bq/kg for ^{134}Cs (n = 3) and 24 ± 42 Bq/kg (n = 87) for ^{137}Cs with a maximum of over 300 Bq/kg. Radiostrontium was found in 67 samples: 5.2 ± 4.5 Bq/kg of ^{90}Sr [53].

Recently, our focus was on tree bark monitoring. The gamma-ray analysis of the tree bark of 26 different trees gave a mean of 6.7 ± 18 Bq/kg ^{137}Cs.

Compared to the undisturbed situation on a country site, variability in soil and vegetation in a city and surroundings is quite dominant. Nonetheless, we think that tree bark monitoring is comparable with soil monitoring and can give relevant contamination data for emergency cases. The uptake mechanism for radionuclides in mosses is quite different to that of trees and of the deposition on soils. Mosses do not have roots; they incorporate contamination mainly through the air. Contamination is deposited on tree leaves. Trees accumulate contamination through their roots and also later by the deposited leaves (litter-fall) [54].

Despite their great variability in moss species and the difficulties in the determination of age, mosses can be monitoring plants for radioactive fallout, when carefully normalised (**Figure 5**). In Basel, radiocaesium and radioiodine (^{131}I) were detectable in the first rainfall in April 2011, after the catastrophe at Fukushima-Dai-ichi. Here, over 9500 km away from Japan, the fallout could also be detected in moss, grass and soil and even in cow's milk [55].

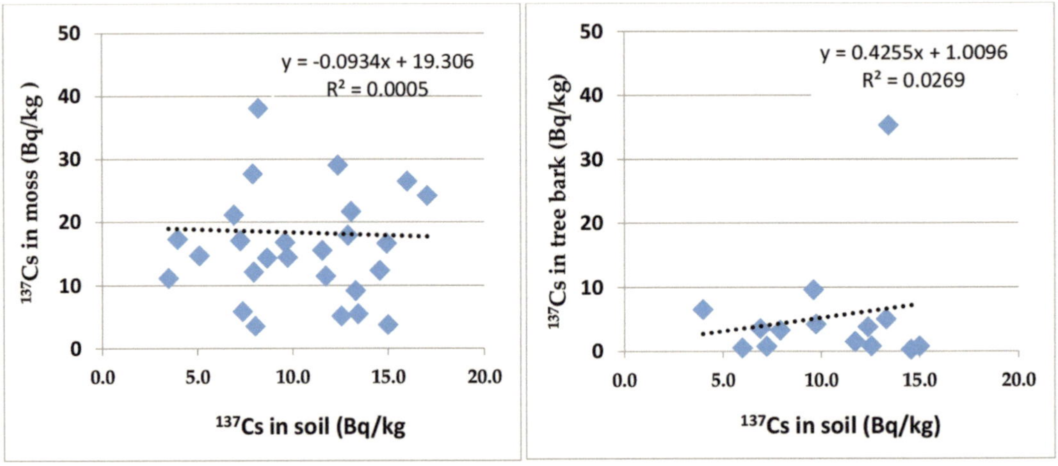

Figure 5. Correlations of radiocaesium between moss, tree bark and the corresponding soil activity.

5. Gamma nuclides in food

5.1. Radio contamination of food

Whereas food can be contaminated just after its release with fallout of short-lived radionuclides for a short period, contamination with long-lived radionuclides from global fallout and the Chernobyl catastrophe remain. The contamination of food by the Chernobyl fallout was reduced within 2 years concerning the short-lived radionuclides, such as ^{131}I, ^{132}I, ^{134}Cs, whereas long-lived radionuclides still persist (see **Figure 3**) in the soil and are transferred to crops and grass (feed for cows). The typical tracer food for this is milk. We recently published a review of radioactivity monitoring in Switzerland over the last 35 years [56]. We compared the contamination level of food categories with artificial radionuclides. In the time span 1990–2015, some moderate contamination of some food categories was noted. Special cases were hazelnuts and tea from Turkey. As some regions of that country were affected by the fallout from Chernobyl, food imports may contain higher levels of Radiocaesium (0.1–30 Bq/kg). Tea contained up to 100 Bq/kg radiocaesium and 2–40 Bq/kg of radiostrontium. The latter may also originate from global fallout. Until the present, most affected food from the fallout of Chernobyl concerns wild grown mushrooms, wild grown berries and game (especially wild boars). Even today, violations are noted for wild boars from Bavaria, Southern Germany and Southern Switzerland [57] (**Table 4**).

Food category	Origin	Radiocaesium	Radiostrontium
Hazelnuts (2005–2013)	Turkey	0.1–16 (n = 96)	n.a.
Milk (2004–2016)	Switzerland	<0.05–22 (n = 177)	0.01–26 (n = 194)
Baby food (2006–2016)	Switzerland, Europe	<0.05–0.5 (n = 71)	<0.1–0.33 (n = 71)
Mushrooms	Europe	<1–1'320 (n = 248)	n.a.
Vegetables (2006–2014)	Europe	<0.1–0.5 (n = 88)	<0.1–0.5
Wild grown beeries[1]	East Europe	0.1–170 (n = 123)	<0.05–60 (n = 57)
Game	Europe	<1–1250	n.a.
Wild boar (2013–2014)	South Switzerland	1250 (n = 41)	n.a.
Wild boar (2014) [58]	Canton of Zurich	<0.2–388 (n = 28)	n.a.
Fish (2011–2015)	Pacific Ocean	<0.1–0.7 (n = 52)[2]	0.1–0.4 (n = 7)
Flour, bred (2006–2011)	Switzerland	<0.1–5	n.a.
Honey (2004–2012)	Europe	<0.2–51.4	<0.05–2
Tea[2]	Japan, other countries	<0.5–258.2	<0.1–57

[1] Including blue berries and chest nuts.
[2] Fish and Japanese tea contains also ^{134}Cs from local fallout of the Fukushima-Daiji accidents.
n.a.: not analysed.

Table 4. Some food categories which are still contaminated with global fallout and/or fallout from NPP's accidents in Chernobyl and Fukushima.

The Office of Public Health estimates the total ingested dose to about 0.3–0.4 mSv/year. The main contribution comes from potassium-40 (^{40}K; 0.2 mSv/year) and from natural radionuclides of the uranium and thorium decay series. The remaining contamination from bomb fallout is less than 0.1 mSv/year [39].

5.2. Healing earths

Siliceous earths are widely used in the food industries as a food supplement. They incorporate foreign atoms in the crystal lattice, such as radionuclides of the natural decay series of uranium and thorium.

In 2008, siliceous earth products on the Swiss market were analysed with γ-spectrometry. In two products, the threshold value for natural radionuclides of group 2[1] was exceeded (>50 Bq/kg). Furthermore, by regular consumption of one product from California, USA, the annual dose would reach half of the permitted yearly dose of 1 mSv. Consequently, this product was withdrawn from the Swiss market. In 2010, the reinspection of healing earths showed that two products from one producer in Germany slightly exceeded the limit value according to higher levels of ^{226}Ra and ^{228}Ra. The annual dose from the consumption of these products would lead to 0.1 mSv/year. Therefore, healing earths and silica-based chemicals used in food industry and in chemical laboratories remain a source of natural radionuclides [59].

6. Gamma spectrometry as an important analytical tool for emergency cases with ionising radiation

The instrumentation of an emergency A-Laboratory depends on the required detection devices for the analysis of the fallout from nuclear bombs or from nuclear power plants. **Table 5** gives a short survey of some expected fission and activation products. In used reactor fuel, more than 200 radionuclides are present [60] (**Figure 6**). The Institut de Radiophysique Appliqué (IRA) prepared simulated gamma spectra 2 and 11 days after release from an NPP. After 2 days, the gamma analysis was very complex and contained more than 270 gamma emission lines. At day 11 after the release, the number of gamma lines was reduced remarkably. Not all expected radionuclides were detectable due to low activity. For mother/daughter nuclide pairs, attention has to be paid to when the half-life of the daughter is shorter than that of the mother. After seven half-lives of the daughter, the two nuclides are in equilibrium, so one has to calculate the activity of the daughter using the half-life of the mother nuclide (e.g. ^{132}Te (77.5 h) and daughter ^{132}I (2.3 h)) [61].

As we see, gamma spectrometry is the most important instrumentation for an emergency case with ionising radiation. Only for a few radionuclides α-spectrometry (Pu-isotopes) and β-spectrometry (^{3}H, ^{14}C, ^{89}Sr, ^{90}Sr) have to be available.

The main task is to analyse environmental samples, such as soil, vegetation, fallout, air and food samples, such as vegetables and fruit grown outdoors. The pathways are air/fallout, rain/washout and water (rivers, lakes). Later on, collection and analysis of sediment, grass and soil samples will follow. The most important/affected food is milk and milk products, baby food, outdoor-grown vegetables, meat and game, fruit (including hazelnuts) and cereals. According to our experience in 1986, vegetables were the most affected (by radioiodine and radiocaesium). Special focus should be set on baby food and human milk (including analysis for radiostrontium) [55].

[1]Ordinance on Contaminants and Constituents in Food: natural radionuclides group 2: the sum of activities of 226Ra, 228Ra, 230Th, 232Th and 231Pa.

Radionuclide	Half-lives	detection	Radionuclide	Half-lives	detection
Noble gases			**Non-volatile nuclides**		
^{85m}Kr	4.4 h	γ	^{60}Co	5.3 a	Γ
^{87}Kr	1.3 h	γ	^{54}Mn	312 d	γ
$^{88}Kr/^{88}Rb$	2.8 h/50 d	γ	^{65}Zn	244 d	γ
^{133}Xe	125 h	γ	$^{97}Zr/^{97}Nb/^{97m}Nb$	17 h/1.2 h/53 s	γ
^{133m}Xe	53 h	γ	^{89}Sr	50 d	β[1]
			$^{90}Sr/^{90}Y$	28.6 a/64 h	β[1]
Halogens			$^{91}Sr/^{91}Y$	9.5 h/59 d	γ
^{131}I	192 h	γ	$^{92}Sr/^{92}Y$	2.7 h/3.5 h	γ
^{132}I	2.3 h	γ	$^{95}Zr/^{95}Nb$	65 d/35 d	γ
$^{133}I/^{133}Xe$	21 h	γ	^{60}Co	5.3 a	γ
^{134}I	0.9 h	γ	$^{99}Mo/^{99m}Tc$	66 h/6 h	γ
$^{135}I/^{135m}Xe$	6.6 h/0.3 h	γ	$^{97}Zr/^{97}Nb$	17 h/1.2 h	γ
^{82}Br	35 h	γ	$^{127}Sb/^{127}Te$	3.9 d/9.4 h	γ
			$^{129}Sb/^{129}Te$	4.4 h/70 m	γ
Volatile nuclides			$^{127}Sn/^{127}Sb$	2.1h/3.9d	γ
$^{129}Te/^{129}I$	70 m/12.4 h	γ	$^{140}Ba/^{140}La$	306 h/40 h	γ
$^{129m}Te/^{129}Te$	802 h/1.2 h	γ	^{141}Ce	33 d	γ
$^{131m}Te/^{131}Te$	30 h/0.4 h	γ	$^{143}Ce/^{143}Pr$	33 h/14 d	γ
$^{132}Te/^{132}I$	78 h/2.3 h	γ	$^{144}Ce/^{144}Pr$	285 d/17 m	γ
$^{134}Te/^{134}I$	42 m/52 m	γ	^{149}Pm	53 h	γ
$^{133m}Te/^{133}I$	55 m/21 h	γ	$^{235}U/^{231}Th$ etc.	7×10^8 a	γ
$^{103}Ru/^{103m}Rh$	941 h/56 m	γ	$^{241}Pu/^{241}Am$	14.35 a/432 a	γ
$^{106}Ru/^{106}Rh$	374 d30 s	γ	$^{239}Np/^{239}Pu$ etc.	2.4 d	γ
^{134}Cs	2.1 a	γ	$^{242}Cm/^{238}Pu$ etc.	163 d	α
$^{136}Cs/^{136m}Ba$	316 h/	γ	$^{238}Pu/^{234}U$ etc.	88 a	α
$^{137}Cs/^{137m}Ba$	30a/2.5 m	γ	$^{239}Pu/^{235}U$ etc.	2.4×10^4 a	γ
^{3}H	12.4 a	β[1]	$^{240}Pu/^{236}U$ etc.	6.6×10^3 a	α
^{14}C	5730 a	β[1]	$^{242}Pu/^{238}U$ etc.	3.7×10^5 a	α

After Refs. [60–62].
[1]Analysis has to be performed with β-spectrometry.

Table 5. Extract of possible fission and activation products released at an NPP accident or from a bomb.

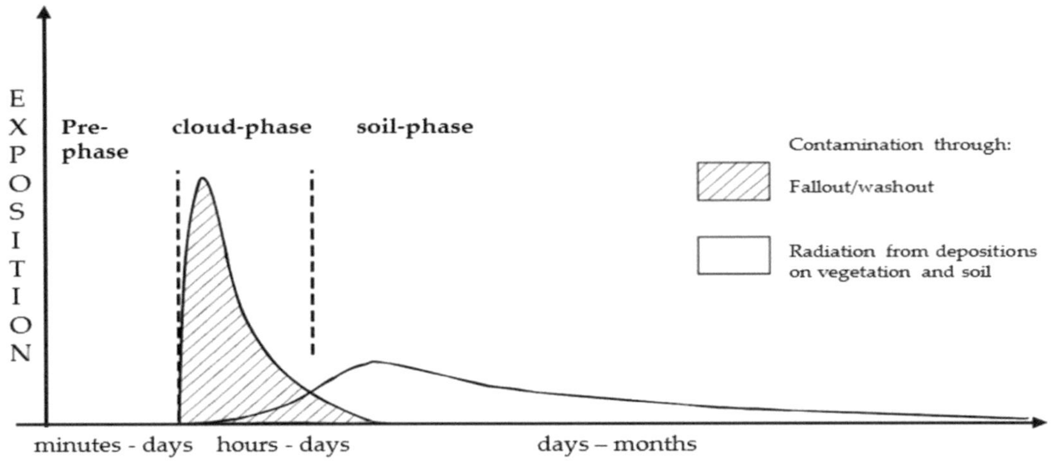

Figure 6. Process of a release of radioactive material to the environment by an accident or bomb (after Ref. [64]).

In Switzerland, the National Emergency Operations Centre (NEOC) coordinates these emergency investigations. They give the order to collect samples for the estimation of the contamination/radiation level outdoor and for controlling the level of the contamination of food. The local authorities then take action (e.g. banning severely contaminated food categories). An initial focus is set on the drinking water quality. In Basel, drinking water is produced from groundwater, which is enriched with river water by soil filtration. It is important to prevent the entry of contaminated river water into this filtration system. After an earthquake, when the drinking water production site is no longer operable, there exist plans for emergency supply of the public with pumped water from the ground and from rivers. This requires efficient and rapid analysis systems. With a gamma analysis of a 1 L water sample in a Marinelli-beaker, it is possible to restrict the analysis time to 15 min for the examination of threshold values. ^{3}H and ^{90}Sr have to be analysed by β-spectrometry. With an extraction/scintillation-method, we are able to obtain results for both radionuclides within 2 hours [63].

Acknowledgements

My special thanks go to my former collaborator, Matthias Stöckli. I benefited greatly from his great expertise in radiation detection, especially in Neutron Activation Analysis. Furthermore, I express my thanks to my collaborators Franziska Kammerer and Michael Wagmann for their permanent support. Finally, I wish to thank Major Franz Näf from the civil protection of Basel City for his engagement during the annual emergency exercises that are mentioned in Section 6.

Author details

Markus R. Zehringer

Address all correspondence to: markus.zehringer@bs.ch

Head of Radiation Laboratory, State-Laboratory of Basel-City, Basel, Switzerland

References

[1] Knoll G. Radiation Detection and Measurement. 3rd ed. New York: John Wiley & Sons; 1999. ISBN 0-471-81504-7, p. 387ff.

[2] Gilmore G, Hemingway J. Practical Gamma-ray Spectrometry. New York: John Wiley & Sons; 1995. ISBN: 0-471-95150-1.

[3] L'Annunziata M, editor. Handbook of Radioactivity Analysis. 2nd ed. San Diego: Academic Press; 2003. ISBN 0-12-436603-1, p. 239–294.

[4] Debertin K, Helmer R. Gamma- and X-Ray Spectrometry with Semiconductor Detectors. New York: Elsevier; 1988. ISBN: 0-444-871071.

[5] Ortec. [Internet]; Available at: http://www.ortec-online.com/

[6] Canberra. [Internet]; Available at: http://www.canberra.com/products/detectors/ [Accessed: 2016-08-02].

[7] Institute of Metrology Czec [Internet]; Available at: https://www.cmi.cz/?language=en [Accessed: 2016-08-02].

[8] Interwinner by ITECH INSTRUMENTS [Internet]; Available at: http://www.itech-instruments.com/ [Accessed: 2016-08-02].

[9] Vojtyla P, Beer J, Stavina, P. Experimental and simulated cosmic muon background of a Ge spectrometer equipped with a top side anticoincidence proportional chamber. Nucl Instr Methods Phys Res B. 1994; **86**: 380–386.

[10] Seo B, Lee K, Yoon Y, Lee D. Direct and precise determination of environmental radionuclides in solid materials using a modified Marinelli beaker and a HPGe detector. Fresenius J Anal Chem. 2001; **370**:264–269.

[11] Valkovic V. Determination of radionuclides in environmental samples. In: Barcelo D, editor. Environmental Analysis: Techniques, Applications and Quality. Amsterdam: Elsevier Science Publishers; 1993, p. 311–356.

[12] Wallbrink P, Walling D, He Q. Radionuclide measurement using HPGE gamma spectrometry. In: Zapata F, editor. Handbook for the Assessment of Soil Erosion and Sedimentation Using Environmental Radionuclides. Amsterdam: Springer; 2003. ISBN 978-0-306-48054-6. p. 67–87.

[13] Lederer M, Shirley V. Table of Isotopes. 7th ed. New York: Wiley & Sons; 1978. ISBN: 0-471-04179-3.

[14] Erdtmann G and Soyka W. The Gamma Rays of the Radionuclides. Weinheim: Verlag Chemie; 1979. ISBN: 3-527-25816-7.

[15] Wahl W. α—β– –γ-Table of Commonly Used Radionuclides, Version 2.1; 1995. D83722 Schliersee.

[16] Firestone R, Shirley V. Table of Isotopes. 8th ed. New York: Wiley & Sons; 1999. ISBN: 978-0-471-35633-2

[17] Legrand J, Perolat J, Lagourtine F, Le Gallic Y. Table of Radionuclides. Atomic Energy Agency, Gif-sur-Yvettes, France 1957.

[18] Amiel S (Ed.). Nondestructive Activation Analysis, Amsterdam: Elsevier; 1981. ISBN: 0-444-41942-X.

[19] Parry S. Handbook of Neutron Activation Analysis. Woking: Viridian Publishing; 2003. ISBN: 0-9544891-1-X.

[20] Zehringer M, Stöckli M. Determination of total bromine residue in food and non-food samples. Chimia 2004;**59**:112.

[21] The Federal Department of Home Affairs. Ordinance on Contaminants and Constituents in Food; 1995.

[22] Zehringer M, Testa G, Jourdan J. Determination of total Iodine content of food with means of Neutron Activation Analysis (NAA). Proceedings of the Swiss Food Science Meeting (SFSM '13); 2013; Neuchatel.

[23] Bhagat P et al. Estimation of iodine in food, food products and salt using ENAA. Food Chem. 2009; **115**: 706–710.

[24] Frey T. Detection and quantification of brominated flame retarding agents in plastic materials. [Master's thesis]. Aarau; 1999.

[25] Kuhn E, Frey T, Arnet R, Känzig A. Brominated flame retarding agents in plastic products on the Swiss market. In: Federal Office of Environment (Ed.) Umweltmaterialien; 2004:**189**.

[26] Wegmann L, Werfeli M, Bachmann R, Tremp J, Figueiredo V. Brominated flame retarding agents in plastics. A tentative investigation on the Swiss Market, Amt für Umweltschutz und Energie. Liestal; 1999.

[27] Zehringer M et al. Neutron Activation Analysis (NAA)—another approach to uranium and thorium analysis in environmental samples. Chimia 2013; **67**:828.

[28] Mahaffey J. Atomic Accidents. A History of Nuclear Meltdowns and Disasters. New York: Pegasus Books; 2014. ISBN: 978-1-60598-680-7.

[29] Valerius-Mahler C. Radiation—the two faces of radioactivity. Pharmazie-Historisches Museum der Universität Basel; 2014, 67–87.

[30] Pratzel H, Deetjen P. Radon Application at Sanitariums. Geretsried: I.S.M.H Verlag; 1997.

[31] Pöschl M, Nollet L, editors. Radionuclide Concentrations in Food and the Environment. New York: CRC Taylor & Francis; 2007. ISBN: 0-8493-3594-9.

[32] Bünger T et al. Radioactivity in water—an actual overview. Strahlenschutzpraxis. 2014;**1**: 3–28.

[33] Zehringer M. Radiological investigation of a radium drinking device. In: annual report of the state-laboratory Basel-City. Basel; 2006. p. 173–175.

[34] Zehringer M. Wall tiles and other radioactive objects. In: Annual Report of the State-Laboratory Basel-City. Basel; 2006. p. 165–167.

[35] Marzocchi O. Analysis of a radioactive lens and of the correlated UV-light bleaching properties. Strahlenschutzpraxis 2015;**2**: 46–49.

[36] Marfeld A. Modern Electronics and Practice of Electro Technique. Berlin: Safari-Verlag. 1971. p. 210.

[37] Buckley D, Belanger R, Martin P, Nicholaw K, Swenson J (Eds.). Environmental assessment of consumer products containing radioactive material. Report of the U.S. nuclear regulatory commission NUREG/CR-1775; 1980.

[38] Izrael Y. Radioactive Fallout after Nuclear Explositions and Accidents. New York: Elsevier; 2002, eBook ISBN: 9780080540238. p. 102.

[39] Federal Office of Public Health. Environmental Radioactivity and Radiation Doses in Switzerland. Bern: BAG [internet]; Available from: http://www.bag.admin.ch/themen/strahlung/12128/12242/index.html?lang=de [Accessed: 2016-08-02].

[40] National Emergency Operations Centre (NAZ) [Internet]; Available at: https://www.naz.ch/de/aktuell/messwerte.html [Accessed: 2016-08-02].

[41] Swiss Federal Nuclear Safety Inspectorate (ENSI) [Internet]; Available at: https://www.ensi.ch/de/notfallschutz/messnetz-maduk/ [Accessed: 2016-08-02].

[42] Federal Office of Public Health (BAG) [Internet]; Available at: http://www.bag.admin.ch/themen/strahlung/12128/12266/12268/index.html?lang=de [Accessed: 2016-08-02].

[43] Federal Office of Public Health (BAG) [Internet]; Available at: http://wp.radenviro.ch/?p=891&lang=de [Accessed: 2016-08-02].

[44] Geering J, Friedli C, Carlone F, Lerch P. Analysis of strontium-90. In: Federal Office of Public Health, editor. Environmental Radioactivity and Radiation Doses in Switzerland. Bern: BAG; 1991. ch.B.3.6.7.

[45] Corcho-Alvarado, J et al. Long-term behaviour of 90Sr and ^{137}Cs in the environment. Case studies in Switzerland. J Environ. Rad. 2016;**160**: 54–63.

[46] Zehringer M. Waste water monitoring of the ARA Basel. In: annual report of the state-laboratory Basel-City Basel; 2015. p. 129–131.

[47] Zehringer M. Waste water monitoring at KVA Basel. In: annual report of the state-laboratory Basel-City. Basel; 2015. p. 132–134.

[48] Rumpel N. Balancing of short-lived radionuclides for medical use in Basel-City [scholarly paper]. Basel: state-laboratory Basel-City. Basel 2015.

[49] Kammer F, Rumpel N, Wagmann M, Zehringer M. Monitoring of Radionuclides for medical use in Basel. Proceedings of the GDCH-Wissenschaftsforum, Dresden; 2015.

[50] Rumpel N, Kammerer F, Wagmann M, Zehringer M. Gamma ray spectrometry of sewer sludge—a useful tool for the identification of emissions sources in a city. Chimia 2015;**69**: 301.

[51] Zehringer M. Radioactivity monitoring of the river Rhine. In: Annual Report of the State-Laboratory Basel-City. Basel; 2015. p. 134–138.

[52] Abraham J. Study of the behavior of radionuclides at an infiltration site for drinking water production. Basel: State-Laboratory Basel-City. Basel; 2016.

[53] Meyer J. Monitoring of moss in Basel-City for the investigation of radioactive fallout [scholarly paper]. Oberwil: State Grammar School Oberwil; 2014.

[54] Agapkina G et al. Dynamics of Chernobyl-fallout radionuclides in soil solutions of forest ecosystems. Chemosphere 1998;**36**: 1125–1130.

[55] Zehringer M. Radioactivity in food and other objects from Japan. In: Annual Report of the State-Laboratory Basel-City. Basel; 2011. p. 29–31.

[56] Zehringer M. Radioactivity in Food: Experiences of the Food Control Authority of Basel since the Chernobyl Accident. In: Monteiro W. editor. Radiation Effects in Materials. InTech Open. 2016. ISBN: 978-953-51-2438-2.

[57] Palacios M, Estier S, Ferreri G. On the trace of ^{137}Cs in wild boars in Ticino In: Federal Office of Public Health, editor. Environmental Radioactivity and Radiation Doses in Switzerland. Bern: BAG; 2013. p. 100–101.

[58] Swiss wild boar meat: radioactivity and heavy metal content. In: Annual Report of the State-Laboratory of Zurich. Zurich; 2014. p. 71–72.

[59] Zehringer M. Radionuclides in siliceous and healing earths. In: Annual Report of the State-Laboratory Basel-City. Basel; 2008. p. 79–82.

[60] Choppin G, Rydberg J. Nuclear Chemistry—Theory and Applications. New York: Pergamon Press; 1980. ISBN: 0-08-023826-2. p.593 Apendix H.

[61] Buchillier T, Baillat C, Laedermann J, Leupin A. Rapport sur l'exercice d'analyse de spectre de centrale nucléaire 2013; Rapport interne de l'Institut de Radiophysique Appliqué, Lausanne, Switzerland, 2015.

[62] Demongeot S et al. A practical guide for laboratories measuring radionuclides at post-accidental situations. IRSN (Institut de Radioprotection et de Sureté Nucléaire), Rapport No 2011-02, 2011.

[63] Zehringer M, Abraham J, Kammerer F, Syla V, Wagmann M. Fast survey of Radiostrontium after an Emergency Incident involving Ionizing Radiation. Chimia 2016;**70**: 816.

[64] Federal Commission for ABC-protection. Concept for emergency protection in the vincinity of NPPs, KomABC 2006-03-D; Berne: 2006.

All references concerning the reports of the State Laboratory Basel City are on the internet, available at: http://www.kantonslabor.bs.ch/berichte.html.

Chapter 2

Dead Time in the Gamma-Ray Spectrometry

Salih Mustafa Karabıdak

Additional information is available at the end of the chapter

http://dx.doi.org/10.5772/67083

Abstract

A review of studies of the dead time correction on gamma-ray spectroscopy is presented. Compensate for counting losses due to system dead time is a vital step for quantitative and qualitative analysis. The gamma-ray spectroscopy system consisting of electronic devices are used for detection of radiation due to gamma rays. The dead time of the spectroscopy system is based on time limitations of these electronic devices. Firstly, a new model for determination of this electronic dead time is proposed. Secondly, two alternative methods suggested for the correction of this electronic dead-time losse.

Keywords: gamma rays, gamma-rays detectors, semiconductor detectors, dead time, counting rate, peaking time

1. Introduction

Development of a gamma spectrometer for each energy region of the electromagnetic radiation has progressed in parallel with the development of experimental tools. The first and rough detectors often just used to determine the presence of radiation. The second-generation radiation detectors used to determine the radiation intensity, but with only a very small part of information on its energy. The last types of radiation detectors measure the intensity as a function of the photon energy in addition to the determination of the presence of radiation.

After the first observation of the gamma rays with photographic plates, advances related to measuring in this field began with the development of various types of gas-filled counters at the beginning of 1908 [1]. The counters that are compared to the photographic detection process allowed the experimenter to obtain a more accurate quantitative measure of the radiation as well as determining the presence of the radiation. The proportional counters did allow one to obtain energy spectra for gamma rays whose energies were low enough to interact primarily by the photoelectric effect and where the secondary electrons produced by these interactions could be completely stopped in the gas volume. However, generally these detectors are only used in determining the number of events that occur in the counter and not to measure directly the energy of the incoming photons [2].

The main improvement in determining the quantity measurement of gamma rays began with the development of NaI (Tl) detectors in about 1948. These detectors can supply energy spectra over a wide energy range. After a certain period of development, detectors consist of crystals with sufficiently large size to allow high absorption rates even above 1 MeV photons energy were produced. The main advantages of these detectors include their relatively fine resolution, the good physical and chemical stability of the crystal material used and their high relative yield. As these detectors have a good resolution, photon energies are well separated and it allows the observation of different energy photon peaks [2].

In 1962, semiconductor Ge (Li) detectors were manufactured [3]. As these detectors can be made from many different semiconductor materials, these are used as photon detectors as well as nuclear-charged particles detectors. To collect the secondary charges efficiently, these detectors need to be made of single crystals of a pure material. Due to difficulties in producing single crystals other than germanium, Ge (Li) detectors have, so far, been successfully used as high-resolution photon detectors of significant size. These detectors have a high resolution. The most serious drawback of semiconductor detectors is the need to keep them cold, generally at liquid nitrogen temperature.

In the coming years, there have been numerous studies on similar detectors of high atomic number. Mayer [4] proposed several detectors made from bicomponent material. Sakai [5] had worked on semiconductor detectors made from bicomponent material such as GaAs, CdTe and HgI_2 to make measurements at room temperature. However, there is not much use of these detectors up to now because of the relatively low yield due to their small surface area, low resolution and expensive production methods and techniques.

Characteristics of an ideal detector for gamma spectrometer can be expressed as follows [6]:

1. Output pulse number should be proportional to the gamma-ray energy.
2. It should have a good efficiency (that should be a high absorption coefficient).
3. To collect the detector signals, it should have an easy mechanism.
4. It should have good energy resolution.
5. It should have good stability over time, temperature and operating parameters.
6. It should be conferred at a reasonable cost.
7. It should have a reasonable size.

http://dx.doi.org/10.5772/67083

2. General characteristics of photon detectors

Modern photon detectors used in determining the radiation and in measuring the quantity of radiation run based on a series of joint steps. These steps are the same in almost all species and types of detectors. In this section, characteristics of gas detectors and of semiconductor detectors commonly used in the determination of radiation and in the measurement of the quantity of radiation are discussed. These detectors are used to count electrons, heavy charged particles and photons. We will only focus on their use as a photon detector. A photon detector operates over the following principles [2]:

1. First, the conversion of the photon energy to kinetic energy of electrons (or positrons) by photoelectric absorption, Compton scattering or pair production.
2. Second, the production of electron-ion pairs, electron-holes pairs, or excited molecular states by these electrons.
3. Third, the collection and measurement of the charge carriers or the light emitted during the deexcitation of the molecular states.

A photon spectrum released by a source is usually consists of monoenergetic photons group. The detector converts such a line spectrum into a combination of lines and continuous components. Detectors can be used to determine the energies and intensities of the original photon as long as these lines are observable. However, if the lines are lost in the associated continuity it is usually not possible to determine these quantities. The ability of the detector to produce peaks and lines for monoenergetic photons is characterized by the peak efficiency and peak width. The peak width in the gamma-ray spectrometer is usually expressed as the full width at half maximum (FWHM) in terms of keV. In addition, FWHM referred to as the resolving power of the detector. The peak efficiency of the detector is the ratio of the number counts in the peak corresponding to the absorption of all the photon energy (so in the full energy peak) to the number of photons of that energy emitted by the source. Both the peak and the peak efficiency are functions of the photon energy [2].

3. Gamma-ray spectrometry system components

The magnitude of the pulses from the gamma ray is equal to the magnitude of electrical charge, which is proportional to the amount of absorbed gamma-ray energy by gamma-ray detectors. The function of the electronic system is to collect these electrical charges, to measure the amount of electrical charge and to store these information. The electronic system for a gamma-ray detector spectrometer is shown schematically in **Figure 1**.

The gamma-ray detector system consists of a detector bias supply, preamplifier, amplifier, analog-to-digital converter (ADC), multichannel analyzer (MCA), a data storage device (computer and spectrum analysis program), a pulse generator and oscilloscope if desired. Pulse generator is used in the spectroscopy system, which does not contain an electronic circuit with functions such as a base line restorer or a pileup rejector. Detector bias supply generates

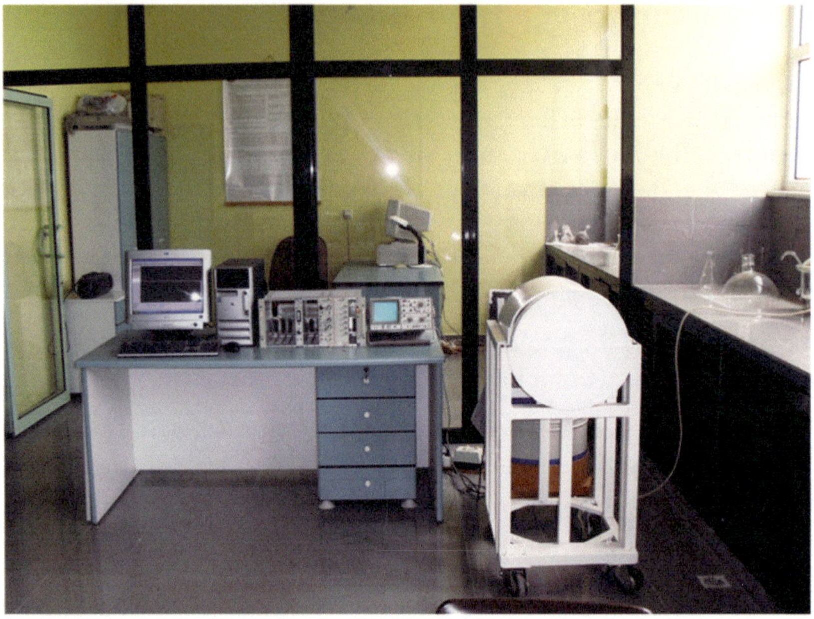

Figure 1. General appearance of the gamma-ray spectrometry system.

electric field, this produced electric field sweeps the electron-hole pairs toward detector contacts. These swept electron-hole pairs are collected by preamplifier. The collected electron-hole pairs in the preamplifier are converted to a voltage pulse with a field effect transistor (FET). The amplifier changes the shape of the voltage pulse. This shape of the pulse increases linearly with the size of the incoming pulse. The analog-to-digital converter converts from the analog structure to the digital structure. The multichannel analyzer (MCA) shorts the pulses according to pulse height in addition; MCA counts the number of pulses in the individual pulse height ranges. Computers are used to check the measurements and to record the spectrum in modern gamma-ray spectrometers. The advantage of such systems is providing great convenience to users in fulfilling their various data-analysis calculations during and after measurement. During the measurement, the location and area of the peak of interest via a program used can be determined on screen, therefore, the collection rate of counting and the identity of radionuclide can be determined.

4. The dead-time detection methods

The main reason for the counting and pileup losses is the dead time of the gamma spectrometry system. In order to fulfill the necessary corrections primarily related to these losses, it is to be first determined the dead time of the gamma-rays spectrometry. The dead time is associated with limited time features known as constant separation time of electronic circuits of the gamma-ray spectrometry. It has been generally accepted that the direct measurement of the dead time because of the pulse processing conditions in the gamma-ray spectrometry may change or not is accurately known. Traditional dead-time measurement techniques are based on the fact that the observed count rate varies nonlinearly with the true counting rate. Therefore, by assuming that one of the specific models is applicable and by measuring the observed rate for at least two different true counting rates that differ by a known ratio, the dead time can be calculated [7].

4.1. Two sources method

This method is based on the observation of the counting rate from two individual sources and in combination of these two sources. Because of the counting losses are nonlinear, the observed counting rate resulting from the two sources combined is less than the sum of the observed counting rates resulting from each individual source counting. Thus, the dead time can be calculated from this mismatch [7]. Considering two sources, such as A and B, Prussin [8] gave this relationship:

$$n_A + n_B = n_{AB} + n_{BG} \tag{1}$$

Where n_A, n_B, n_{AB} and n_{BG} show the observed counting rate of A, B, $A + B$ (A plus B) source and background, respectively. Considering the counting rate correction in the case of nonparalyazable and zero background, general solution to this statement is as follows:

$$T_D = \frac{n_A n_B - [n_A n_B (n_{AB} - n_A)(n_{AB} - n_B)]^{1/2}}{n_A n_B n_{AB}} \tag{2}$$

4.2. Decaying source method

This method can be applied if the source is a short-lived radioisotope. In this method, the decay constant of the radioactive source must be known. The dead time due to the observed count rate resulting from exponential decay of the source is determined. A graph is plotted using the observed counting rate and the decay constant of the source and dead time can be determined with the help of this graph [7]. A general approach to obtain accurate count rates in this method is that: first, the net count rate versus time is plotted on the semilogarithmic graph paper, to obtain a linear curve. This curve is fitted linearly by the least square method. The resulting fit equation is as follows [7, 8]:

$$n(t) = n_0 e^{-\lambda t} + n_{BG} \tag{3}$$

where n_0 is the true rate at the beginning of the measurement and λ is the decay constant of the particular isotope used for the measurement. If the paralyzable model used for this case, the solution can be given as follows:

$$\lambda t + \ln m = -n_0 \tau e^{-\lambda t} + \ln n_0 \tag{4}$$

4.3. The electronic dead time

In this recently proposed method, the dead time is a result of the electronic components of the gamma-ray spectrometry [9]. Charged particles that are generated by incident radiation within the detector crystal are transported by an electric field to the detector electrodes [9]. The production and collection of the charged particle are subjected to random statistical variations, which depend on the incident energy and the detector medium. An intrinsic resolution limitation exists in the process of converting the incident radiation to an electrical signal [9, 10]. The output signal undergoes various processing steps in order to be correctly acquired and analyzed in semiconductor X or gamma-ray detectors. The time required to collect the charged particles produced by the incident radiation is important in many applications. If the collection time is not sufficiently short compared with the peaking time of the amplifier, a loss in the recovered signal amplitude occurs [9, 11]. The charge collection time depends on the detector geometry, medium, electric field and location of the interaction within the detector active volume [9].

An optimized spectrometer system provides the best energy resolution obtainable within a given set of experimental constraints. System optimization requires the proper selection of equipment and knowledge of the compromise of resolution and count rate performance in any system [9]. The detector and preamplifier combination is the most critical component of the system electronics. The best amplifier cannot compensate for poor signal-to-noise or count rate limitations caused by improper selection of the system front end. Selection of the proper amplifier will enhance the performance of the good detector and preamplifier combination. The source and detector interaction, detector and preamplifier combination, pulse processor shaping and the system count rate determine the system resolution [9, 10].

4.3.1. Determination of the electric dead time

The relationship between the minimum resolving time, peaking time and overall pulse width is given by the following equation [9, 10]:

$$T_R \geq \frac{T_W}{T_P} - 1 \tag{5}$$

where T_R is minimum resolving time, T_W is overall pulse width and T_P is peaking time of the amplifier. Overall pulse widths in response to possible peaking times of the amplifier can be measurement using the oscilloscope. A graph is plotted using the measurement overall pulse widths and the possible peaking time of the amplifier (see **Figure 2**). The relationship between overall pulse width and peaking time can be determined with the help of this graph. The fitting equation of data in **Figure 2** is,

$$T_W = B_3 T_P^3 + B_2 T_P^2 + B_1 T_P + A \tag{6}$$

where A, B_1, B_2 and B_3 are coefficients of the equation fitted [9]. The minimum resolving times in response to possible peaking times of the amplifier were calculated using Eq. (5) and the minimum resolving time versus peaking time is plotted in **Figure 3** [9].

The fitting equation of the data in **Figure 3** is [9],

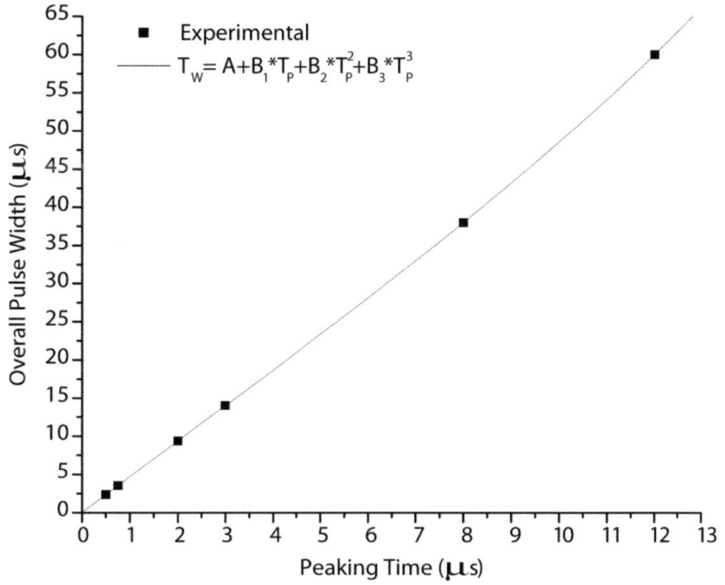

Figure 2. Change in the overall pulse width with peaking time [9].

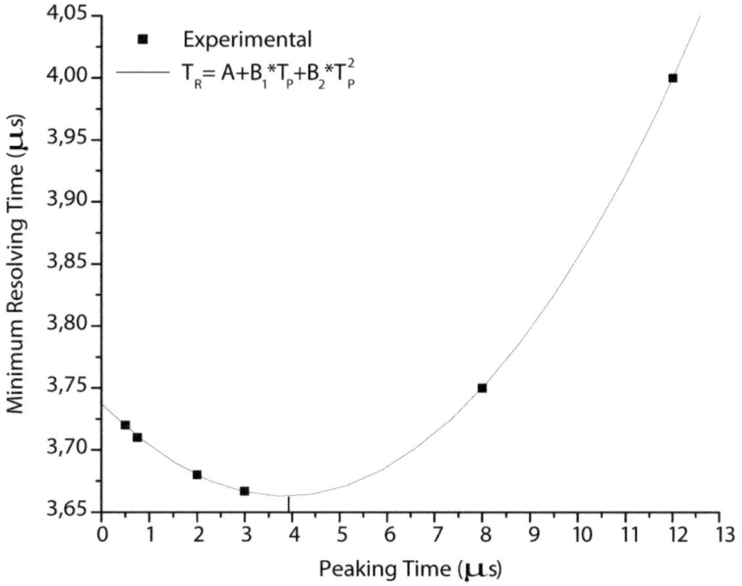

Figure 3. Minimum resolving time versus peaking time [9].

$$T_R = B_2 T_P^2 + B_1 T_P + A \tag{7}$$

The T_D effective system dead time can fall into one of the following categories [9]:

$$T_R > 1.5\mu s + T_C \tag{8}$$

or

$$T_R < 1.5\mu s + T_C \tag{9}$$

where T_C is conversion time of the analog-to-digital converter (ADC). If the minimum resolving time of the pileup rejector is greater than 1.5 μs + T_C, then the system dead time is simply [9]

$$T_D = T_P + T_R \tag{10}$$

Otherwise, the system dead time is

$$T_D = T_P + 1.5\mu s + T_C \tag{11}$$

The ADC conversion time for the relevant energy lines can be calculated [12] by using,

$$T_C = \frac{E}{\Delta E} T_{\text{Clock}} \tag{12}$$

where E is energy line, ΔE is energy per channel and T_{Clock} is the amplifier operation frequency. Considering Eqs. (7) and (10)–(12), the system dead time can be written as [9]:

$$T_D = B_2 T_P^2 + (B_1 + 1) T_P + A \tag{13}$$

or

$$T_D = T_P + 1.5\mu s + \frac{E}{\Delta E} T_{Clock} \tag{14}$$

5. Some models for the dead-time correction

Illustration of the effect on the counting rate of the dead time can be done with the detector; a pulse-forming network consists of a square wave output with constant τ length (amplifier and analog-to-digital converter (ADC) and a counter (multichannel analyzer (MCA)). This time τ is variable due to caused dead time of the ADC.

5.1. Paralyzable (extended) model

One model of dead time behavior of gamma-ray spectroscopy system is paralyzable (extended) response. This model has come into common usage. The model represents idealized behavior. True events that occur during the dead time are lost and assumed to have no effect, whatsoever on the behavior of the detector. True events that occur during the dead time, however, although still not recorded as counts, are assumed to extend the dead time by another period τ following the lost events. This method can be expressed as follows [7]:

$$m = ne^{-n\tau} \tag{15}$$

where m is the recorded count rate, n is the true count rate and τ is the system dead time.

For the so-called paralyzable (extendable) systems, the dead time is extended by starting from the last arrival time. Pulse pileup can also be interpreted as a kind of pulse loss of the paralyzable type [13]. Consecutive pulses falling within a time interval peaking time, T_P, are treated as pileup and excluded from the spectrum, as a pileup rejector (PUR) [14]. To generate second output pulse without a time interval of at least s between two consecutive true events is not possible in the paralyzable model. In this model, the recovery of the electronic device is further extended during the respond time s to an initial event for an additional time s by some additional true events, which occur before the full recovery has taken place [15].

5.2. Non-paralyzable (nonextended) model

Another model of dead-time behavior of gamma-ray spectroscopy system is nonparalyzable (nonextended) response. This model has come into common usage. The model represents idealized behavior. A fixed time τ is assumed to follow each true event that occurs during the live time of the detector. True events that occur during the dead time are lost and assumed to have no effect, whatsoever on the behavior of the detector. True events that occur during the dead time, however, although still not recorded as counts, are assumed to not-extend the dead time by another period τ following the lost event. This method can be expressed as follows [7]:

$$n = \frac{m}{1-m\tau} \tag{16}$$

where m is the recorded count rate, n is the true count rate and τ represents the system dead time.

In nuclear spectrometry measurements, the pulse loss is traditionally related to the nonparalyzable (nonextendable) dead time per incoming pulse caused by an ADC during pulse processing. Nonparalyzable means that the dead time period is not prolonged by a new pulse arriving during that time [14]. In the nonparalyzable model, recovery period of electronic device is not affected by events that have come into being during the s dead time [16].

5.3. Live-time correction model

On the assumption that all the MCA dead-time losses are manifested through the input rate, the accumulated live time takes as the proper counting time of the measurement should be both necessary and sufficient. The live time is the actual time during which the system is open and available for collecting counts. Thus, the counts within a spectral peak must be divided by live time (T_{live}, in seconds) to obtain the counts per second. This model is well in low count rate situations [6].

5.4. Gedcke-hale model

This model is a development of the previous one, which also compensates for losses due to leading edge pileup in the amplifier and is favored by EG&G Ortec. It predicts a correction based on Poisson statistics. This method can be roughly expressed as follows [6]:

$$\text{Counts to memory} = \text{input pulses to amplifier}\frac{T_{\text{Live}}}{T_{\text{Real}}} \tag{17}$$

Where T_{live} is the live time and T_{real} is the real time of the counting system.

5.5. Pulser Model

In this model, the pulser generates constant amplitude pulses that are similar to the pulses output from the detector. These pulses are sent to preamplifier. Through the electronic components of the gamma-ray spectrometry, the produced pulses are transported and stored in the memory as a pulser peak. It is a fair assumption that the fractional losses sustained by the pulser counts are the same as those sustained by the gamma-ray derived counts. Thus, dead-time losses in the gamma spectrometry may be allowed for by multiplying the gamma peak areas by the following simple ratio [6]:

$$\frac{\text{pulses produced by pulser}}{\text{Pulser counts in pulser peak}} \tag{18}$$

5.6. Loss-free counting model

There is a number of methods can be grouped under this general heading. All of them make use of a subsidiary circuit to monitor the instantaneous count rate and based on that generate a weighting factor, n. Thus, the high instantaneous count rate is reflected by a high count in a channel in the spectrum that is appropriate at that time [6].

The Harms procedure was a pioneering effort. It counts those pulses that are presented for processing but which are rejected because the system is busy. The number of discarded pulses in such a scale is read and used to weight the next real event. However, the ADC processing time is not necessarily the major problem at high count rates as losses due to pulse pileup can dominate [6].

5.7. Zero dead-time counting model

Changing the dead time problems of gamma-ray spectrometry is a source of a well-known error. The dead-time correction under such conditions may be true if changing the dead time dominantly affected measured radioisotopes [17, 18]. Zero dead-time losses correction is almost the same as the loss-free counting. The difference between them is completely quantified of zero dead-time correction.

5.8. Integral dead-time correction model

The major constraint in the gamma-ray spectrometers is both the time required to collect the charge produced by ionizing radiation in the active detector volume and subsequently the pulse processing by the electronics [14, 19]. In this major constraint of all gamma-ray spectrometers, there is a minimum amount of time called dead time of the spectrometer system. During this dead time, the system cannot respond to other incoming photons and these events cannot be counted and thus can be lost [14]. One of the major problems confronting the user of gamma-ray spectrometer is to correct the results for counts lost due to the spectrometer dead time. This problem can be solved automatically by carrying out all counting runs for a known or measured total instrument live time rather than for real time [14, 20].

When count rate is kept nearly constant, counting losses due to dead time can be corrected by a simple formulae for both types of nonextendable and extendable dead times [14, 21]. However, when the count rate changes or fluctuates significantly, the correction based upon mathematical means becomes difficult and complex. In order to overcome this problem, a method was demonstrated by Kawada [14, 22], which allows compensating the dead time effects automatically at every moment during the counting experiment. In this method, pulses whose number is equivalent to the dead-time losses were generated as random coincidence pulses using a gating technique in a first-order approximation and added to the output pulse train after delay [14, 23].

The problem of varying dead time is a well-known source of error in nuclear spectrometry measurement. In this case, dead-time corrections can only be accurate if the varying dead time is dominantly caused by a radiation source. Solutions have been offered in several forms: dead time stabilization [14, 24–26] solves the problem at the cost of a fixed, perhaps unnecessary, dead time and resulting loss of counting efficiency [17]. However, results of the other dead-time correction models have not been forthcoming for counting losses due to the system dead time. It is a reliable estimation of the original count rate or of the number of original events for the interval of time considered. To do this, we need either an accurate value for the average loss per dead time or a method that allows us to arrive at individual corrections. An analytic correction method was developed instead of using a pulser or a radioactive source. This proposed model by Karabıdak et al. [14] is based on a measuring principle on the total live time.

5.8.1. Background of the integral dead-time correction model

In Galushka's study [14, 27], a method is described for restoring dead-time losses in real time so that at the output of a counter, constructed according to this new scheme, one obtains directly by the number of events expected in the absence of dead time. This is accomplished by inserting additional pulses into actual series of registered events. In such a situation, let the observed sequences of events be characterized by the arrival times T_0, T_1, T_2 ... [14, 28]. Consecutive arrivals are separated at least by the peaking time, T_P, applied. If we put $T_0 = 0$, then pulse number, k, occurs at the instant [14]:

$$\begin{aligned} T_k &= T_1 + T_2 + \ldots + T_k \\ &= (\tau + \delta_1) + (\tau + \delta_2) + \ldots + (\tau + \delta_k) \\ &= k\tau + \sum_{j=1}^{k} \delta_j \, for \, k \geq 1 \end{aligned} \tag{19}$$

where δj is the width of each pulse, which is separated from each other by steady dead time arising from peaking time of the amplifier. However, peaking time is important for calculations of counting losses due to the dead time of the system. Therefore, the dead time of the system is determined by adding the peaking time to the ADC converting time. To determine the dead time of the system, minimum resolving time should be ascertained first [14].

For counting losses due to systems dead time, both approaches are possible. Traditional correction formulae were used for the first method: they are based on the observed count rate and are applied at the end of a measurement period [14]. On the contrary, methods of a second type work in a different way by instantly correcting or compensating for losses, apparently without requiring knowledge of the measurement or calculate count rate. In the second method, it is possible to estimate the probability of losing a specific number k of counts in a dead time of length TD. Since we deal with a Poisson process, this probability is given by Refs. [14, 23, 28]:

$$P_k = \frac{(nT_D)^k}{k!} e^{-nT_D} \tag{20}$$

where n is the count rate in each channel. The expected counting losses due to each dead time are given as follows [14]:

$$L = \sum_{k=1}^{\infty} kP_k = e^{-nT_D} \sum_{k} \frac{(nT_D)^k}{(k-1)!} = nT_D \tag{21}$$

Thus, correction counting is given by Ref. [14]:

$$C_C = Count + L \tag{22}$$

5.9. Differential dead-time correction model

Recording two pulses apart as two different events at almost all detectors systems requires to be separated from another pulse. This situation needs the minimum time interval [15]. The minimum time interval based on electronic devices using the counting system is usually called the dead time of the counting system. This is generally determined by pileup reject time, paralyzable or nonparalyzable system dead time or a combination of these mechanisms [13]. The photons arriving to the detector at the dead time period are not being counted. Thus, count rate, which is expressed as the count of per unit time, decreases [15].

The paralyzable and nonparalyzable models are assumed to express the idealized behavior. Every true event came into being during live time of detector was assumed to occur in a stable s dead time at every two models [7, 15]. However, this situation is only valid in the dead time depending on peaking time of amplifier. Nevertheless, a counting system can be containing analog-to-digital converter (ADC) which determines the energy value of pulse. The dead time of such a counting system is variable with ADC conversion time [14]. Therefore, fulfilled corrections, which are taken into account for a fixed dead time, cannot be realistic. In addition, modern counting systems consist of electronic devices, which contain paralyzable (amplifier), nonparalyzable (ADC) and pileup reject (amplifier) [15, 29]. The paralyzable and nonparalyzable models predict the same first-order losses and differ only when true event rates are high. These models have two extreme idealized system behaviors and real counting systems often display a behavior that is intermediate between these extremes. The detailed behavior of a specific counting system may depend on physical processes taking place in the detector, delay introduced by the pulse processing and recording electronics [15].

In medium and high-count rate events, both of the two models are not applicable. The corrections, which are done by these models, are problematic because of the limitations expressed below. The troubling aspect of nonparalyzable model is the singularity at $m\tau - 1$ and the fact that a maximum observed counting rate of $1/\tau$ is approached in the limit as n approaches infinity. In the paralyzable model, the observed counting rate becomes zero at high-count rate. In addition, it should be noted that this model could not be explicitly solved for n_0. Nevertheless, this model solves a transcendental equation to obtain the true counting rate. In addition, the observed counting rate is either double valued or does not exist above a maximum value given by $\exp(-1/\tau)$ [14, 30].

This model can be applied to the counting systems at which the system dead time is not predominant on count rates. That is, this method adequately corrects counting lost at steady counting rate. In addition, the dead-time or count-rate corrections based on live time can be ideal in the count rates which are not predominate at the system dead time [15]. In addition, on a mathematical essence, the principle of the live time is an integral mathematics. The integral mathematics is correct if applied only to stationary Poisson processes (invariable in time). It should be noted that time-invariant Poisson processes are valid in experimental studies with radionuclides having long half-lives. The current study includes count-rate corrections based on differential mathematics and the proposed model in this study is ideal in the count systems at which the system dead time is predominant on count rates. Differential mathematics is also correctly applicable to Poisson process changing in time [15].

5.9.1. *Background of the differential dead-time correction model*

Kurbatov et al. [15, 31] proposed a correction method that included to a statistical approach for Geiger—Muller counter. All of photons emitted from source are assumed to be caught by the detector and transmitted to counting system without loss. Let $P(t)$ be the probability that a photon is emitted from a source in the interval $(t-\tau, t)$. Let $a(t)dt$ be the probability that a photon is caught by detector and transmitted to counting system during the interval $(t, t + dt)$. The fact that $P(t)$ is a continuous function of t is used here. In order that a photon can be caught by the detector and sent to counting system, it is necessary and sufficient that; (i) a photon is sent counting system from detector in the time interval (t, $t + dt$) and (ii) no counting take place in the time interval ($t - \tau$, t). Since these are independent events, the realization probability of one counting in the time dt becomes. Then, since only one counting can occur in the interval ($t - \tau$, t), the probability of one counting in that interval is [15, 31]:

$$P(t) = \int_{t-\tau}^{t} [1-P(x)]a(x)dx \tag{23}$$

When $1-Q(x) = P(x)$ conversion is taken into consideration, the following equation can be written as:

$$Q(t) = 1-\int_{t-\tau}^{t} Q(x)a(x)dx \tag{24}$$

If t is too large of τ, $a(t)$ is independent of t. For a preparation whose decay constant is λ, containing N_0 atoms at time zero, f approaches to while N_0 increases [15, 32, 33]. Then, the expected number of photons reaching the detector in t time without dead time (the $P(x)$ probability equal zero) is given by Ref. [15]:

$$C(t) = \int_{0}^{t} N_0\lambda e^{-\lambda t}dt = N_0(1-e^{-\lambda t}) \tag{25}$$

Considering these inferences, the differential correction model can be created in the following way: λ on Eq. (17) is known as "decay constant" and is defined as the number of decay particles per second. In addition, unit of λ is 1/second. On the other hand, the number of particles counted per second by the detector defines counting rate. Therefore, unit of the counting rate is 1/second. Decay constant and counting rate are equivalent from the perspective of analogical. Therefore, the counting system consisting of only amplifier or ADC or both amplifier and ADC can be considered as a decaying source. In that case, n_0 maximum number of counting rate of the amplifier or the ADC or both the amplifier and ADC can be compared with N_0, the number of particles at $t = 0$ in a radioactive source. Thus, Eq. (25) may be rearranged when n_0 instead of N_0 and λ [15]:

$$n(t) = \int_{0}^{t} n_0 n_0 e^{-n_0 t}dt = n_0(1-e^{-n_0 t}) \tag{26}$$

Where $n(t)$ is the count rate recorded by the counting system which consists of a detector, an amplifier and ADC.

For a compound system, observed counting rate in each channel of X or gamma-ray spectrum is given by Karabıdak et al. [14]. Karabıdak and Çevik [15] calculated the dead time for amplifier, ADC, or both amplifier and ADC. Thus, for a compound system, counting rate correction, (or true counting rate) corresponding to each channel can be satisfied by:

$$n_0 = \frac{1}{T_D} \ln\left(\frac{1}{1-m \cdot T_D}\right) \tag{27}$$

In this case, corrected count in each channel at a counting period in a spectrum is given by Ref. [15]:

$$C_C = n_0 T_{Real} \tag{28}$$

where T_{Real} is the real time of the counting system.

6. Conclusion

A practical method to determine the dead time is proposed by Karabıdak and colleagues [9]. In other methods to determine the dead time, the dead time due to the amplifier and ADC is determined separately. In addition, to compensate for the counting losses kept constant dead time is fulfilled by considering a fixed dead time. Wherein, the dead time in this method is obtained at the same time for both the amplifier and ADC. An effective way of decreasing counting losses is by decreasing the system dead time in quantitative and qualitative analysis. Thus, the dead time is that variables in the counting process are fulfilled and the counting losses for a unified system are easily compensated. Because the system dead time is linked to the amplifier peaking time, the amplifier peaking time can be set to lower and optimum values.

The integral dead-time correction is effective both for low count rates and for medium count rates. It is possible to observe the dead time contributions due to ADC conversion time and peaking time in this model. Thus, counting losses correction arising from system dead time can be made for demanded situation (nonparalyzable or paralyzable or both). The dead time of the counting system was determined with an analytic formula. Counting losses occurring during this dead time were compensated for by considering uncorrected spectra obeying the Poisson behavior. This new method adequately corrects counting loss at steady counting rate [14].

The differential dead-time correction is effective both for medium count rates and for high-count rates. Output count rates of the only ADC and both the amplifier and ADC are the same. In addition, output-counting rates of the only amplifier are higher than others output counting rates. Moreover, increasing peaking time of the amplifier increases both the dead time and the counting losses. In addition, this model easily determined the relationship between the output counting rates and input counting rates. According to the dead time of the counting system, the amplifier or ADC or both the amplifier and ADC may be taken into account up to a certain limit value. This limit value of such electronic devices represents a saturation point. This saturation point is determined by the size of the dead time of the counting system. This is a result of Poisson statistics. Since the dead time is a function of the peaking time of the amplifier and photon energy, this saturation point is directly related to them. Therefore, while low peaking time determines low dead time, low dead time determines the high saturation point (or counting rate) [15].

The dead time increases as long as the counting rate increases. This increase in dead time becomes stable at around 1 s after a certain point. This fixed point corresponds to the counting rate at saturation point of the electronic system at which the detector is not considered to be receiving the photons [15].

Author details

Salih Mustafa Karabıdak

Address all correspondence to: smkarabidak@gumushane.edu.tr

Department of Physics Engineering, Faculty of Engineering and Natural Sciences, Gümüşhane University, Gümüşhane, Turkey

References

[1] Rutherford E, Geiger, H. An electrical method of counting the number of α-particles from radio-active substances. Proc. R. Soc. Lond. A 1908;**81**:141–161.

[2] Debertin K, Helmer RG. Gamma and x-ray spectrometry with semiconductor detectors. Amsterdam: Elsevier; 1988. 402 p.

[3] Pell EM. Effect of Li-B ion pairing on Li+ ion drift in Si. J. Appl. Phys. 1960;**31**:1675–1680. DOI: 10.1063/1.1735914

[4] Mayer JW. Semiconductor detectors for nuclear spectrometry II. Nucl. Instr. Meth. 1966;**43**:55–64. DOI: 10.1016/0029-554X(66)90532-5

[5] Sakai E. Present status of room temperature semiconductor detectors. Nucl. Instr. Meth. 1982;**196**:121–130. DOI: 10.1016/0029-554X(82)90626-7

[6] Gilmore G, Hemingway JD. Practical gamma-ray spectrometry. 2nd ed. Chichester: John Wiley and Sons Inc.; 2008. 390 p.

[7] Knoll GF. Radiation detection and measurement. 3rd ed. New York: Jhon Wiley and Sons Inc.; 2000. 796 p.

[8] Prussin SG. Prospects for near State-of-the art analysis of complex semiconductor spectra in the small laboratory. Nucl. Instr. Meth. 1982;**193**:121–128. DOI: 10.1016/0029-554X(82) 90685-1

[9] Karabıdak SM, Kaya S, Çevik U, Çelik A. Determination of proper peaking time for Ultra-LEGe detector. Radiat. Measur. 2011;**46**:446–450. DOI: 10.1016/j.radmeas.2011.01.023

[10] Tennelec Instruction Manual TC 244 Amplifier. Oxford Instruments Inc. Analytical Systems Division. Nuclear Measurements Group. USA, 1986.

[11] Gerardi G, Abbene L, Manna AL, Fauci F, Raso G. Digital filtering and analysis for a semiconductor x-ray detector data acquisition. Nucl. Instr. Meth. A 2006;**571**:378–380. DOI: 10.1016/j.nima.2006.10.113

[12] Spieler H. Introduction to radiation detectors and electronics [Internet]. 2009. Available from: http://www-physics.lbl.gov/~spieler/physics_198_notes_1999/index.html [Accessed: 14-02-2009]

[13] Pommé S. Time distortion of a Poisson process and its effect on experimental uncertainty. Appl. Radiat. Isot. 1998;**49**:1213–1218. DOI: 10.1016/S0969-8043(97)10048-3

[14] Karabıdak SM, Çevik U, Kaya S. A new method to compensate for counting losses due to system dead time. Nucl. Instr. Meth. Phys. Res. A 2009;**603**:361–364. DOI: 10.1016/j.nima.2009.02.005

[15] Karabıdak SM, Çevik U. Decaying source model: alternative approach to determination of true counting rates x and gamma ray counting systems. Radiat. Measur. 2009;**58**:18–23. DOI: 10.1016/j.radmeas.2013.07.009

[16] Evans RD. The atomic nucleus. T M H edition. New Delhi: McGraw-Hill; 1955. 990 p.

[17] Blaauw M, Fleming RF, Keyser R. Digital signal processing and zero dead time counting. J. Radioanal. Nucl. Chem. 2001;**248**:309–313. DOI: 10.1023/A:1010603403264

[18] Blaauw M, Fleming RF. Statistical properties of gamma-ray spectra obtained with loss-free or zero-dead-time counting and ORTEC'S "variance spectrum". Nucl. Instr. Meth. A 2003;**505**:306–310. DOI: 10.1016/S0168-9002(03)01074-X

[19] Pommé S. Experimental test of the "zero dead time" count-loss correction method on a digital gamma-ray spectrometer. Nucl. Instr. Meth. A 2001;**474**:245—252. DOI: 10.1016/S0168-9002(01)00884-1

[20] Deighton MO. Statistical errors arising from use of a gated pulse train for total live time measurement during pulse amplitude analysis. Nucl. Instr. Meth. 1961;**14**:48–52. DOI: 10.1016/0029-554X(61)90051-9

[21] A Handbook of Radioactivity Measurements Procedures. NCRP Report No. 58. National Council on Radiation Protection and Measurements. USA,1985.

[22] Kawada Y. A new method of measuring resolving times of counting system and its application. In Proceeding of the 13th Annual Meeting on Radioisotopes in the Physical Sciences and Industry; Tokyo, Japan; 1976.

[23] Kawada Y, Kobayashi S, Watanabe K, Kawamura T, Hino Y. Automatic compensation of dead time effects. Appl. Radiat. Isot. 1998;**49**:1123–1126. DOI: 10.1016/S0969-8043(97)10031-8

[24] Schönfeld F. Alpha—a computer program for the determination of radioisotopes by least-squares resolution of the gamma-ray spectra. Nucl. Instr. Meth. 1966;**42**:213–218. DOI: 10.1016/0029-554X(66)90188-1

[25] Görner W, Höhnel G. An automatic life time correction in multichannel counting of short-lived nuclides. Nucl. Instr. Meth. 1970;**88**:193–195. DOI: 10.1016/0029-554X(70)90494-5

[26] Wiernik M. Comparison of several methods proposed for correction of dead-time losses in the gamma-ray spectrometry of very short-lived nuclides. Nucl. Instr. Meth. 1971;**95**:13–18. DOI: 10.1016/0029-554X(71)90034-6

[27] Galushka AN. The method of Poisson's fluxes of accidental events registration, communication through Müller JW. 1993.

[28] Müller JW. Some remarks on the Galushka method, Rapport BIPM 93/2. Bureau International des Poids et Mesures. France, 1993..

[29] King SH, Lim CB. Pulse pile-up, dead time, derandomization and count rate capability in scintillation gamma cameras. IEEE Trans. Nucl. Sci. 1985;**32**:807–810. DOI: 10.1109/TNS.1985.4336945

[30] Gardner RP, Liu L. On extending the accurate and useful counting rate range of GM counter detector systems. Appl. Radiat. Isot. 1997;**48**:1605–1615. DOI: 10.1016/S0969-8043(97)00161-9

[31] Kurbatov JD, Mann HB. Correction of G-M counter data, Phys. Rev. 1945;**68**:40–43. DOI: 10.1103/PhysRev.68.40

[32] Ruark AE, Devol L. The general theory of fluctuations in radioactive disintegration. Phys. Rev. 1935;**49**:355–367. DOI: 10.1103/PhysRev.49.355

[33] Schiff LI. Statistical analysis of counter data. Phys. Rev. 1935;**50**:88—96. DOI: 10.1103/PhysRev.50.88

Chapter 3

Pseudo-gamma Spectrometry in Plastic Scintillators

Matthieu Hamel and Frédérick Carrel

Additional information is available at the end of the chapter

http://dx.doi.org/10.5772/67134

Abstract

War against CBRN-E threats needs to continuously develop systems with improved detection efficiency and performances. This topic especially concerns the NR controls for homeland security. This chapter introduces how it is now possible to perform gamma identification using plastic scintillators, which are not conventionally designed for this purpose. Two distinct approaches are discussed: the first one is the chemical modifications of the scintillator itself and the second is introducing new algorithms, Specifically designed for this application.

Keywords: plastic scintillator, gamma spectrometry, metal loading, unfolding, homeland security

1. Introduction

Protection of civilians and facilities against chemical, biological, radiological, nuclear, and explosives (CBRN-E) emerged after 9/11 events and remains since this date of particular importance for countries and states. When combined to the shortage of efficient detectors (e.g., ^{3}He for thermal neutron detection) and a global, worldwide crisis, there is a real need of cheap, yet efficient detectors.

To detect illicit smuggling of gamma-ray sources, a first analysis requires fast gamma spectrometry. The abovementioned equation (increase in terrorists attack + need of cheap detectors for large-scale deployment) naturally leads one to use plastic scintillators (PS) as detectors to be embedded in radiation portal monitors. Despite its high gamma-ray sensitivity, this material is not perfectly suited for this, due mainly to the poor gamma resolution, Precluding therefore any subsequent gamma identification. In the case of gamma-rays emitters indeed, only the Compton continuum and Compton edge are obtained after interaction in the plastic scintillator, and no information of the full energy peak can be observed.

This chapter presents recent improvements concerning potential optimizations for the gamma-ray spectrometry using plastic scintillators as a detector, with a focus on:

- Chemically modifying the nature of the PS. Due to its intrinsic low density and effective atomic number, this family of detectors is not well-suited for gamma-ray spectrometry. However, recent advances in the loading of plastic scintillators with organometallic complexes containing, for example, lead, tin, or bismuth, led to important breakthroughs in this field. As a perspective, nanomaterials are now being included in plastic scintillators, which can afford new and unrevealed specifications. All the advantages and drawbacks of the plastic scintillator loading with organometallics will be fully discussed.
- Spectra classification and deconvolution methods based on specific smart algorithms have shown very promising results to identify gamma isotopes either alone or in mixtures. An important aspect for counterterrorism applications is real-time detection so algorithms which fulfill this requirement are of great interest.

This chapter mostly describes recent advances in the chemical modification of plastic scintillators for Pseudo-gamma spectrometry. The second part introduces dedicated algorithms for the processing of poorly resolved gamma-ray spectra, allowing therefore gamma identification with plastic scintillators [1]. When available, some examples will be provided.

2. Plastic scintillator modifications

2.1. Introduction to plastic scintillators

In a few words, a plastic scintillator is a fluorescent polymer which has the capability to emit photons when excited by an ionizing particle. The discovery and use was reported for the first time by Schorr and Torney as early as 1950 [1], as an extension of previous work on liquid scintillators.

The chemical formulation of a PS is usually composed of an aromatic matrix embedding one or several fluorophores. According to the Förster theory [2] (**Figure 1**), after radiation/matter interaction within the polymer, excitons are transferred from the matrix to the first fluorophore, then to the second fluorophore—so-called the wavelength shifter, allowing the incident energy response to emit close to 420 nm, for two reasons. This wavelength indeed corresponds to the maximum of quantum efficiency of traditional photomultiplier tubes (PMT's) and to the optical transparency domain of the material.

A typical composition of a plastic scintillator is the following (**Figure 2**):

- Matrix, generally based on polystyrene or polyvinyltoluene.
- Primary fluorophore, *p*-terphenyl or 2,5-diphenyloxazole (PPO) are global leaders, but many choices are possible.
- Secondary fluorophore, for example, 1,4-bis(5-phenyl-2-oxazolyl)benzene (POPOP), bis-methylstyrylbenzene (bis-MSB), 9,10-diphenylanthracene (9,10-DPA), etc.

http://dx.doi.org/10.5772/67134

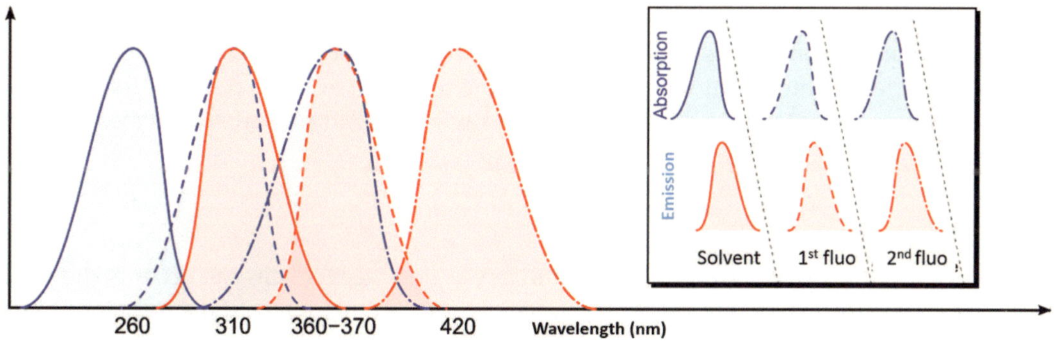

Figure 1. Schematic representation of the Förster theory typically used in liquid or plastic scintillators.

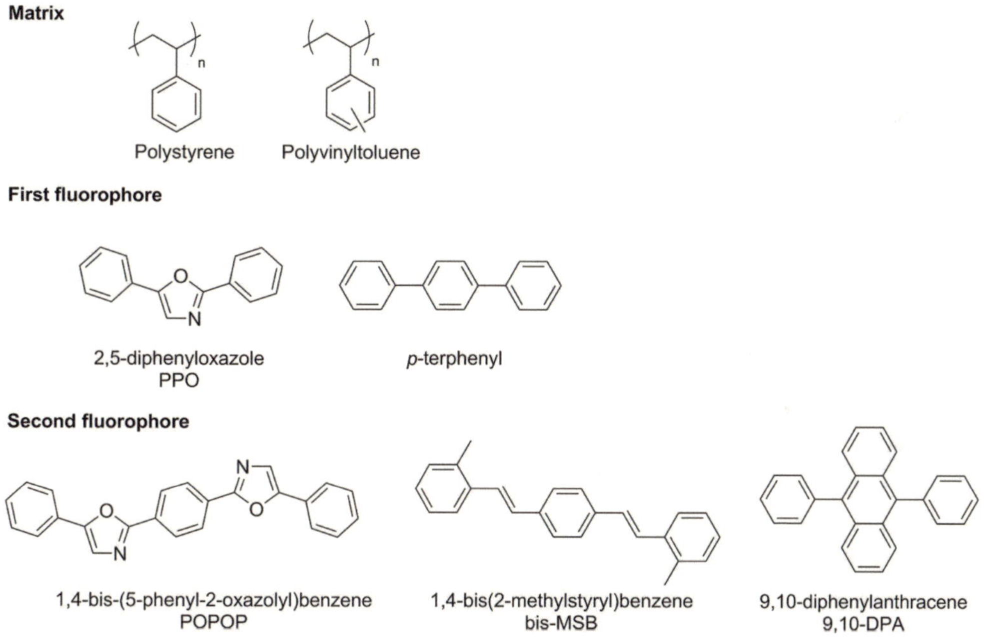

Figure 2. Schematic representations of molecules involved in the preparation of a plastic scintillator.

2.2. Plastic scintillator loading

According to this recipe, one can expect obtaining a regular plastic scintillator, with basic properties. Among other, it is noteworthy at this stage to introduce the scintillation light yield (or light output), which is the quantity of photons delivered by the material when excited by electrons with 1 MeV energy. When a special care is given to afford the plastic scintillator special applications–and this will be the topic of this chapter, it is possible to add various elements, for example, neutron absorbers or organometallics, the latter allowing the Pseudo-gamma spectrometry in plastic scintillators [3].

However, we will see that such loading may affect the light yield, leading to a trade-off between high metal content and detector's performances. In other words, a high loading may increase the effective atomic number (Z_{eff}) of the material, but at the expense of its light output. Also of interest, not only the incident gamma-ray will be fully absorbed, but also the total count rate will benefit from the increase in the Z_{eff}.

2.3. Theory

It is known for a long time that only three main interaction mechanisms can occur at the same time in radiation measurement: Compton scattering, pair production, and the most important for gamma identification is the photoelectric effect.

To identify an incident radionuclide emitting gamma-rays, photoelectric absorption has to be favored. A rough approximation of the photoelectric effect probability is given by $P_{pe} \approx$ constant $\times Z_{eff}^{n}/(E_{\gamma}^{3.5})$, where the exponent n varies between 4 and 5. This equation reveals the main drawback associated with plastic scintillators, namely their low effective atomic number. Calculating Z_{eff} for a PSt + 1.5 wt% *p*-terphenyl + 0.03 wt% POPOP scintillator composition gives a 5.7 value (PSt standing for polystyrene), which is too low to afford a sustainable photoelectric absorption. Thus, Compton scattering will be predominant, and full energy peaks can appear at 100, 300, and 500 keV for compounds with $Z_{eff} \approx 15$, 50, and 70, respectively [4].

To increase both density and Z_{eff}, chemists have turned their attention in loading plastic scintillators with heavy elements, affecting as low as possible the scintillation properties of the detector. This can be illustrated in **Figure 3** where one can see the appearance of the full energy peak on the black signal. To this, four strategies to host a metal inside a polymer matrix are possible. The first strategy to be explained is the most documented and uses organometallics, and the second most used involves nanoparticles. Two other strategies exist, namely dissolution of inorganic compounds and quantum dots, but so far they were not successfully applied to gamma-ray spectrometry; they are therefore not described in this chapter.

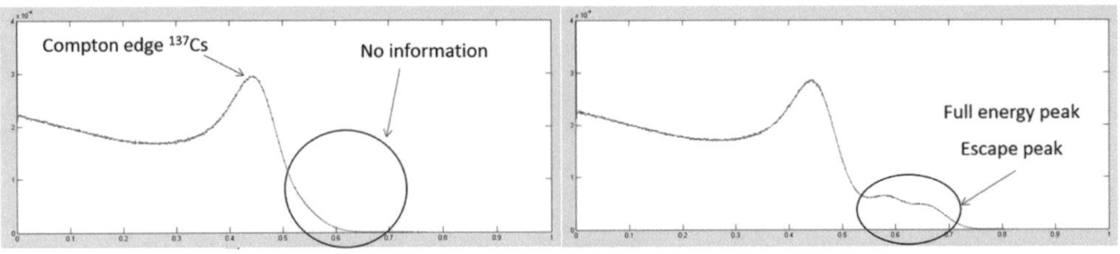

Figure 3. On the left: ^{137}Cs spectrum using a standard plastic scintillator. On the right: ^{137}Cs spectrum using a plastic scintillator doped with 5% lead. MCNPX simulation results. Same behavior expected for bismuth loading.

2.4. Heavy metal loading

2.4.1. Organometallic complex

As already mentioned, the first plastic scintillator ever published was reported in 1950 [1]. Loading such materials with heavy metals took only three years to appear in a publication

[5]. This paved the way on the main characteristics the heavy metal loading has to fulfill: the organometallic complex must be highly soluble in the monomer, it must be stable, and a trade-off must be found between the gain obtained with the loading (herein the increase in the Z_{eff}) and the metal quenching which will affect the light yield.

Basically, an organometallic complex is a molecule embedding a metal core surrounded by one or several organic ligands, their number depending on the valence of the metal. Lead-loaded plastic scintillators are already known and commercially available from several suppliers (**Table 1**). It is shown that the light output dramatically decreases when the metal concentration increases, due (logically) to metal quenching effect. Although not anymore commercially available, a tin-loaded plastic scintillator with metal concentration up to 7 wt% was sold by Nuclear Enterprise under the trade name NE140 [6].

Provider	Reference	[Pb] (wt%)	λ_{em} (nm)	Light yield (ph/MeV)[c]	Light output/ anthracene (%)[c]	Decay time (ns)
Saint-Gobain	BC-452	2, 5 or 10	424	4900	32	2.1
Eljen Technology	EJ-256	1 → 5[a]	425	5200	34	2.1
Amcrys-H	n.d.	12[b]	n.d.	n.d.	n.d.	n.d.
Rexon	RP-452	5	n.d.	5200	34	n.d.

n.d. denotes to not determined.
[a]Eljen Technology can raise the [Pb] up to 10 wt%, but these concentrations *are not recommended* [42].
[b]Also available as tin loading, up to 10 wt%.
[c]Spectroscopic data for 5 wt% loading.

Table 1. Commercial, lead-loaded plastic scintillators.

The choice of both metal and ligand(s) is of particular importance as it will directly affect the scintillation properties of the plastic scintillator. In the literature, the following metals were tested, with more or less popularity, at least in plastic solutions. Thus, Iodine [7] (Z = 53), Holmium [8] (Z = 67), Tantalum [8] (Z = 73), and Mercury [9] (Z = 80) were less studied. An extensive methodological work was performed by Sandler and Tsou concerning the metals from the groups IVA and VA [10]. An important statement was the fact that group VA metals are better quenchers than others of the group IVA.

Diphenylmercury(II) was used in the range 1–10 wt% of metal to load PS. The compound is of particular interest thanks to its high content (56.5%) of metal. Again a strong quenching effect was observed, with a 4-fold decrease in the light output from 1 to 10 wt%. Moreover, diphenylmercury failed to be sensitive to UV light.

Before being radioactive, the two heaviest metals reported in the Mendeleev table are lead and bismuth. Logically, they are also the most studied, along with tin. Thus, after Pichat [5], Hyman [11], Baroni [12], Dannin et al. [13–15] successfully observed for the first time a distinct photopeak emanating from the full energy absorption of the 662 keV, ^{137}Cs gamma-ray. Herein, it is shown that full energy peaks are absent in lead-loaded plastic scintillators and visible in 20–50 wt% of organometallic-loaded materials. Scintillators were composed of 3 wt% 2-(4-biphenylyl)-5-phenyl-1,3,4-oxadiazole (PBD), 0.05 wt% POPOP, and either (4-ethylphenyl)-triphenyltin or *p*-triphenylstyryltin. The best observed photopeak resolution was 13%, obtained with a 30 wt% *p*-triphenylstyryltin (giving 7.86% of metal).

After almost 40 years of inactivity, new developments (mainly driven by the homeland security needs) have recently emerged, for all three recognized metals. Tin has been recently renewed with rationally designed complexes. Despite the lower Z of tin compared to lead or bismuth, the intersystem crossing—and by consequence the metal quenching effect—seems reduced for this metal [16]. An in-depth investigation has been performed in the synthesis of Tin(IV) organometallics in which the steric hindrance pushes back the fluorophore at distances in the range 5–20 Å. These compounds are typically alkyl- or aryl-derivatives of tin, either commercially available or prepared from the Grignard reaction of the organomagnesium halide with tin chloride. A plasticizing effect was observed when loading the material with tetra-(3-phenylpropyl)tin, which led to link the organometallic compound to the matrix *via* a methacrylate moiety. To afford the scintillation properties, 2-(4-*tert*-butylphenyl)-5-(4-biphenylyl)-1,3,4-oxadiazole (butyl-PBD) was used as the dye. Thus, a 6% tin concentration was found as a trade-off between loading ratio and scintillation light yield, providing therefore the presence of photopeaks in the range 14 keV (^{241}Am) to 1274 keV (^{22}Na), with an 11.4% energy resolution at 662 keV. Ultimately, a cost projection was performed with such modified materials. It reveals that a 6 wt% tin-loaded PS could be roughly obtained at a 270 $/kg (chemicals only).

Thanks to its low cost and high Z, lead is still of great interest. As lead(II) organometallics are very polar compounds, most strategies involve the use of polar comonomers to reach high concentrations. Significant improvements were also obtained when adding a polymerizable bond to the organic moiety of the organometallic complex. Thus, lead dimethacrylate ($Pb(MAA)_2$) is a candidate of choice for the 10–40 keV full X-ray absorption in plastic scintillators, allowing a metal loading as high as 27.4 wt% [17]. To allow parasite Cherenkov light effect rejection, this loading was coupled to high wavelength-shifting dyes, leading to 580 nm centered plastic scintillators. Unfortunately, the light yield was rather low (<1000 ph/MeV for 10.9 wt% of lead).

Very recently, a combined fast neutron/gamma discriminating lead-loaded plastic scintillator was reported [18]. This combines an organo-lead compound with PPO at a probable high concentration, this strategy being already known in the field of neutron detection [19]. The resolution is 16% at 662 keV, with a claimed scintillation yield of 9000 ph/MeV.

The last decade has seen many investigations on the synthesis of various bismuth organometallics and their application in plastic scintillator loading. Thanks to the very high solubility of triphenylbismuth in regular monomers such as styrene or vinyltoluene, one can expect elevated bismuth concentrations in the material. To overcome the metal quenching effect, Cherepy et al. proposed an exotic scintillator formulation based on a polyvinylcarbazole (PVK) matrix and a phosphorescent iridium complex. For samples as small as 1 cm^3, the ^{137}Cs pulse height spectrum exhibited several features isolated from bold deconvolutions, such as the full energy peak with a claimed resolution lower than 7%, along with the escape peak due to the ^{209}B Kα X-rays [20]. Bigger volume (48 cm^3) allowed the full absorption of the 1274 keV gamma-ray from ^{22}Na with a 9% resolution [21]. Later on, the PVK matrix was abandoned, presumably due to the difficulty to scale up the material, and triphenylbismuth was substituted by bismuth tripivalate (**Figure 4**) [22]. Loadings of 29 wt% bismuth metal were reached without degraded transparency. A 6 wt% Bi-loaded sample showed a photopeak of ^{137}Cs with a 15% resolution and a light yield close to 3300 ph/MeV.

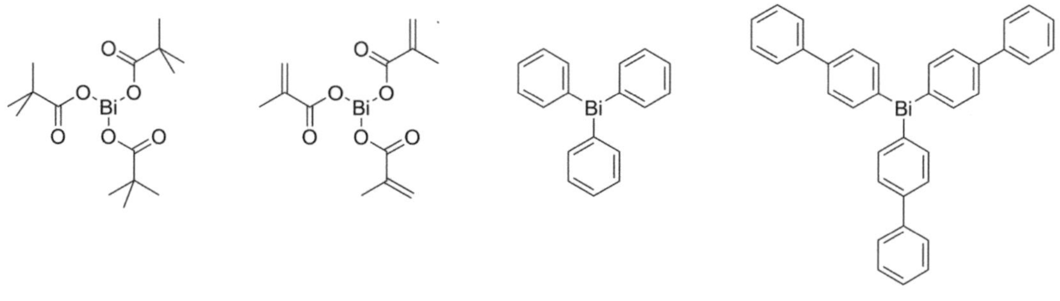

Figure 4. Examples of organo-bismuthine compounds used to load plastic scintillators: bismuth tripivalate, bismuth trimethacrylate, triphenylbismuth, and tris-biphenylbismuth (from left to right).

At the same time, Bertrand et al. designed both alkyl- and aryl-derivatives of bismuth(III) [23]. Bismuth tricarboxylates were easily prepared from the reaction between the carboxylic acid and $BiPh_3$, whereas tris-biphenylbismuth was isolated from the reaction between biphenyl lithium bromide with bismuth(III) chloride. A 6 wt% Bi-loaded sample revealed highly defined full energy peak for ^{57}Co with a 122 keV energy and a light yield close to 4600 ph/MeV. Interestingly, a linear increase in both photoelectric and Compton count rates obtained from the corresponding integration of the region was observed from 0 to 9 wt% of bismuth; the dataset at 11 wt% reveals however a decrease, showing here that the high amount of metal becomes noxious and does not give clear compensation to the loss of scintillation yield. Later on, a 155 cm^3 sized sample containing 5 wt% of bismuth was reported by the same group. A good trend was observed for the full absorption of 67.4 keV X-rays [24]; however, no full energy peak was observed when the scintillator was irradiated with ^{137}Cs gamma source [25].

2.4.2. Nanomaterials

Loading PS with nanomaterials emerge as a new field, and not only for gamma-ray spectrometry, as can be proven for instance with lithium loading for thermal neutron capture [26]. The differences between an organometallic complex and a nanomaterial arise in the sense that the metal core is constituted from a metal salt surrounded by an organic shell, and the global size of the molecule is at the nanometer scale, giving therefore unrevealed feature.

As an application to the gamma-ray spectrometry field, the first example was reported by Cai et al. [27, 28] The composite scintillator was composed of a polyvinyltoluene (PVT) matrix embedding gadolinium oxide (Gd_2O_3) nanocrystals as the gamma sensitizer, along with a rather unusual fluorophore for scintillation purpose, namely 4,7-bis-{2′-9′,9′-bis[(2″-ethylhexyl)-fluorenyl]}-2,1,3-benzothiadiazole (FBtF). FBtF shows an appreciable Stokes shift with a dual excitation maxima at 310 and 420 nm (most probably the $S_0 \rightarrow S_2$ and $S_0 \rightarrow S_1$ transitions, respectively), and an emission maximum around 520 nm. To allow the dispersion of Gd_2O_3 in the composite, it was capped with both oleic acid and oleylamine (**Figure 5**).

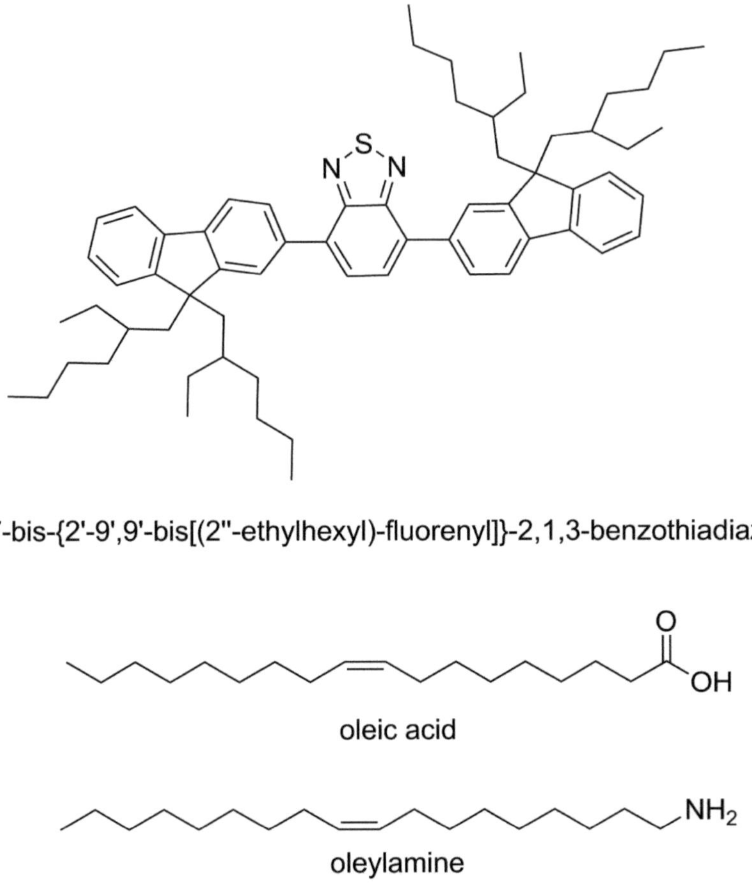

Figure 5. Topological representation of FBtF (top), oleic acid (middle), and oleylamine (bottom).

Thus, small and transparent monoliths (Φ 14 mm, thickness 3 mm) were successfully obtained with a Gd loading as high as 31 wt%. While excited with 662 keV gamma-rays, a full energy peak with an 11.4% energy resolution was observed.

The same group extended this work to hafnium oxide nanoparticles composite scintillators [29]. The nanocomposite monolith of 1 cm diameter and 2 mm thickness shows a full energy photopeak for 662 keV gamma-rays, with the best unfolded full energy peak resolution <8%. Ultimately, ytterbium fluoride nanoparticles with loading as high as 63 wt% were prepared [30]. Composites loaded with 20 wt% of YbF_3 show the full energy peak of ^{137}Cs with an estimated 8600 ph/MeV light yield.

Hafnium-doped organic-inorganic hybrid scintillators were fabricated *via* a sol-gel method [31]. This consists in $Hf_xSi_{1-x}O_2$ obtained from the sol-gel reaction of hafnium oxychloride with phenyltrimethoxysilane dissolved in PSt, with a maximum Hf loading of 10 wt%. Other examples were reported with zirconium oxide nanoparticles loaded in PSt for 67.4 keV X-rays [32]. Loadings up to 30 wt% have been achieved. Significantly higher X-ray excited luminescence was also observed with barium fluoride nanoparticles [33]. Again high loadings (40 wt%) are possible, albeit at the expense of the optical transmission.

3. Algorithms

3.1. Context and motivations

Identification of radionuclides emitting gamma-rays is one of the main challenging topics related to nuclear measurements. This issue is crucial in several fields of nuclear industry. For instance, this step is of great interest in the frame of decommissioning applications or for obtaining a performing storage of nuclear waste packages. Nowadays, identification of radionuclides is considered as a crucial issue for homeland security applications. This important step is for instance required for identifying radioactive material potentially hidden with naturally occurring radioactive materials (NORM).

Gamma-ray spectrometry is the reference line of attack for identifying radionuclides and has been used for decades by scientific and industrial communities. Using appropriate detectors, this technique enables to obtain both qualitative (nature of radionuclides) and quantitative (activity of a given radionuclide) information. However, in the frame of homeland security applications, high detection efficiency is often required and the use of large size detectors became a crucial topic, so plastic scintillators are of great interest. However, due to their intrinsic characteristics, these detectors are not adapted to gamma-ray spectrometry measurements using standard methods.

For several years, alternative methods were developed by different research teams in order to overcome limitations related to detectors having a poor energy resolution. General principle followed by these teams is based on a global analysis of the gamma-ray spectrum and not only on restricted areas of interest in the latter, usually corresponding to a full energy deposition.

3.2. Standard approach for gamma-ray spectrometry analysis and associated limitations

Gamma-ray spectrometry is a technique based on the detection of Gamma-rays emitted by specific radionuclides. This method enables to qualitatively identify radionuclides as well as to quantify their activity. The information of interest is extracted from the full energy peak, which corresponds to a full energy deposition of the incident Gamma-ray. For instance, ^{137}Cs has a characteristic gamma-ray emission at 661.7 keV and the presence of ^{60}Co is associated with the emission of two Gamma-rays at 1173.2 and 1332.5 keV. Using reference database (for instance, ENDF, JEFF, LARA databases [34], etc.), it is possible to identify the radionuclide from a raw gamma spectrum (a calibration step of the detector using well-known sources is required). Concerning the activity information, the latter could be obtained by extracting the net peak area from selected regions in the spectrum. The net peak area corresponds to the gross number of counts in the region of interest minus the background continuum underlying beneath the peak and due to Compton interactions. The ability of discriminating peaks close in energy and extracting their net peak areas are two of the most important features for gamma-ray spectrometry applications.

Detectors used for gamma-ray spectrometry measurements can be segmented into two categories: scintillators and semiconductors. As previously mentioned, the quality of a detector for gamma-ray spectrometry measurements can be evaluated considering two main parameters: its energy resolution (ability to discriminate two peaks in a spectrum close in energy) and its absolute efficiency (number of counts detected in the full energy peak region per emitted Gamma-ray). Both these parameters can significantly change according to the type of the detector. Inorganic scintillators (sodium iodide NaI(Tl), bismuth germanate BGO) have a degraded resolution in comparison with those related to semiconductors (cadmium zinc telluride CZT or high purity germanium HPGe). As an illustration, the standard energy resolution of a NaI(Tl) detector is equal to 7.5% at 661.7 keV, compared to 1 keV for HPGe detector. However, inorganic scintillators often have better detection efficiency. Indeed, due to their low cost, it is possible to produce them in larger dimensions. Some recent technological development, like $LaBr_3$ scintillators, can be considered as a good trade-off for spectrometry measurements (standard energy resolution equal to 3% at 661.7 keV). **Figure 6** compares ^{152}Eu spectra, respectively, obtained using a HPGe detector and a NaI(Tl) scintillator. For the latter, full energy peaks can be distinguished, but the extraction of net peak areas is more complicated than for the analysis of HPGe spectra.

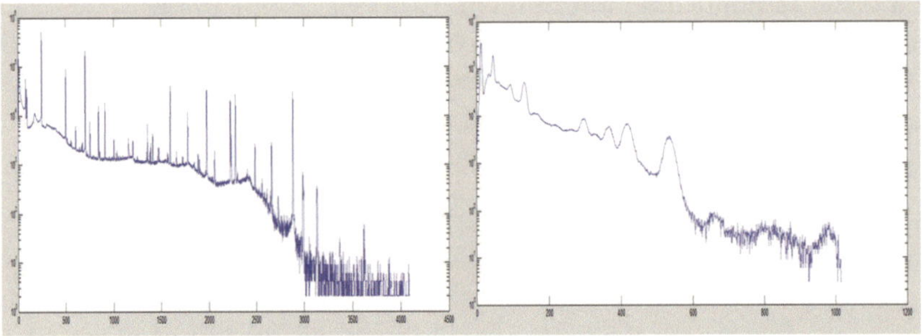

Figure 6. Example of ^{152}Eu experimental spectra using a HPGe semiconductor (on the left) or a NaI(Tl) scintillator (on the right).

On the other side, plastic scintillators are a category of detectors generally devoted to counting applications. For several years, they have been extensively used for homeland security applications. Indeed, productions costs are extremely low, and these detectors can be manufactured in very large dimensions (for instance, an EJ-200 plastic scintillator with dimensions 10 cm × 10 cm × 100 cm, costs 2000 euros per unit). For this reason, there is a lot of interest for extending capacities offered by these detectors, especially adding spectrometric functionalities. However, due to their intrinsic characteristics (plastic scintillators are mainly composed by carbon and hydrogen, corresponding to a Z_{eff} of 5.7 and a density of 1.02) and despite very large dimensions in comparison with other detectors, the probability to have a full energy deposition is very low in the active volume. **Figure 7** illustrates a simulated spectrum of a plastic scintillator in the presence of ^{137}Cs and ^{60}Co sources. The reference Monte Carlo MCNPX code is used to carry out this simulation step [35]. Full energy peaks cannot be identified in the spectrum, but only Compton edges which are often used for energy calibration. For this reason, alternative methods are required to qualitatively and quantitatively process such degraded spectra.

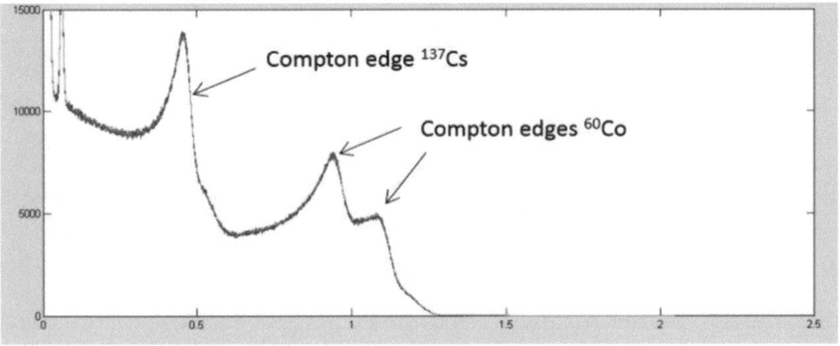

Figure 7. Example of a gamma-ray spectrum obtained with a plastic scintillator (EJ-200 type), ^{137}Cs and ^{60}Co signatures (Monte Carlo simulation using MCNPX).

As it was mentioned in the first part of this chapter, intensive work has been carried out by several research teams for several years in order to enhance the spectrometric performances of plastic scintillators. **Figure 8** illustrates the difference between gamma-ray spectra obtained with a standard plastic scintillator and a 5 wt% lead-loaded plastic scintillator (simulation result using MCNPX). Interesting features due to lead loading are present in the last case (small full energy peak and escape peak). However, a specific spectrum processing is still required to carry out a performing identification step.

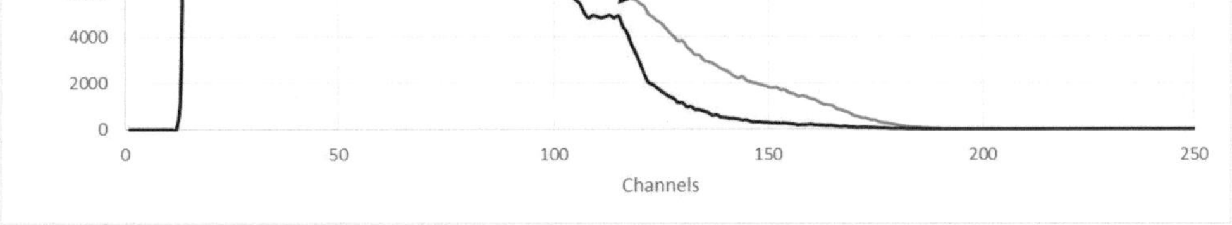

Figure 8. Example of a ^{137}Cs gamma pulse height spectra with a standard (gray line) and a metal-loaded PS allowing the visualization of a full energy peak.

3.3. Analysis of a gamma-ray spectrum as an inverse problem

A potential alternative for analyzing gamma-ray spectra having a poor energy resolution consists of processing the complete spectrum and not only slight regions of interest focused on full energy peaks. In this way, the processing of a gamma-ray spectrum appears as an inverse problem (classical approach for instance for tomographic applications) and can be solved using a specific reconstruction algorithm. Following this methodology, the analysis of a gamma-ray spectrum can be considered under the following matrix form:

$$S = H.A \tag{1}$$

where:

- S is called the signal matrix: matrix of dimensions (*nbe_channels*, 1). The *nbe_channels* parameter corresponds to the number of channels of the gamma-ray spectrum. The S matrix corresponds to the measurement result, that is, the gamma-ray spectrum to be processed.

- A is called the activity matrix: matrix of dimensions (*nbe_incident_energies*, 1). The *nbe_incident_energies* parameter corresponds to the number of incident energies defined by the user and considered during the reconstruction process. These incident energies correspond to the number of voxels (elemental volumes) of a standard emission tomography problem. The A matrix corresponds to the result of the reconstruction.

- H is called the transfer matrix of the problem: matrix of dimensions (*nbe_channels, nbe_incident_energies*). This matrix integrates all detection efficiencies taken into account in the inverse problem. For instance, the element h_{ij} corresponds to the probability that a photon of incident energy equal to j was detected in the channel i of the gamma-ray spectrum. Roughly speaking, it can be seen as a reference database from which the deconvolution step will be carried out.

In this way, the H matrix contains the spectrometric behavior of a detection system for each incident energy defined by the user and considered in the problem. It is important to emphasize that the data processing will be further carried out on this incident energy grid. For this reason, the choice of the grid energy step is a crucial parameter to obtain a performing reconstruction.

Another manner to define the H matrix consists of directly considering gamma-ray signatures of specific radionuclides (for instance, ^{241}Am, ^{137}Cs, ^{60}Co, etc.) and not individual incident gamma-ray energies. In this case, the H matrix has the dimensions (*nbe_channels, nbe_radionuclides)* and the reconstruction step directly enables to obtain the proportion of a given radionuclide in the spectrum. The parameter *nbe_radionuclides* corresponds to the number of radionuclides defined by the end user and considered in the problem.

The definition of the H matrix is one of the most important points of this technique. Two approaches are possible to determine this parameter. The use of Monte Carlo simulations is a first possibility. Several Monte Carlo codes like MCNPX, GEANT4, or TRIPOLI are indeed available and can be considered as performing solutions to determine the behavior of a detector for a given incident energy. The main benefit of this solution concerns its flexibility. Indeed, it is possible to simulate any incident energy or experimental configuration and to minimize in this way experimental constraints due to the calibration step. On the other side, the drawback of this method is mainly related to the accuracy of the Monte Carlo model and its consistency with the experimental behavior of the real detector. A discrepancy between the simulated gamma-ray spectrum and the real behavior of the detector will have a direct impact on the result of the reconstruction. The importance of this impact will be fully related to the agreement between simulated and experimental results. For this reason, an accurate simulation is required for both the detector and its environment which is somehow tricky, especially for the low-energy part of the spectrum (energy range <200 keV, strong impact of Compton effects due to the environment).

The other solution consists of an experimental calibration of the detector itself using reference sources. The number of experimental radionuclides is also reduced in comparison with a Monte Carlo simulation-based methodology, and it will directly define the grid for the reconstruction step (only the database approach can be considered in this way). However, the great benefit of this solution is to obtain the exact experimental behavior of the detector for a given radionuclide, without bias in comparison with the simulation.

The reconstruction step is the second key parameter of this process. First of all, it is important to emphasize that a direct inversion of the formula given previously cannot be considered for solving such problems. Indeed, S or H matrices coefficients have intrinsic statistical uncertainties, and a direct inversion can lead to nonphysical values (for instance, negative activity values, which are of course non-consistent with a physical behavior). Development of specific algorithms has been a topic of interest for the research community for several decades, and many scientific articles were published on this subject. For instance, we can cite the linear regularization system (also named as the Phillips-Twomey approach [36]) or the method based on the maximum of entropy [37] (MEM). One of the most literature-cited methods is based on the approach called maximum likelihood-expectation maximization [38] (ML-EM). In comparison with Phillips-Twomey or MEM methods, ML-EM enables to take into account the Poisson nature of the experimental data given as an input of the problem. An example of applications of this type of algorithm for this current topic can be found in Ref. [39]. Finally, the ML-EM approach can be extended to the Bayesian MAP-EM (maximum a posteriori-expectation maximization) algorithm which introduces an a priori law on the incident energy grid.

Both ML-EM and MAP-EM are iterative techniques, and these algorithms converge to a solution enabling to maximize likelihood. An important point concerns the initial values taken at the beginning of the analysis. A standard procedure consists of considering that all the coefficients of the A matrix have the same values before the first iteration.

3.4. Analysis on simulated and experimental data

Figure 9 illustrates results obtained for the deconvolution of gamma-ray spectra following an inverse problem methodology. A cocktail of radioactive sources is simulated using the Monte Carlo MCNPX code (^{241}Am, ^{137}Cs, ^{60}Co) for a standard plastic scintillator. An ML-EM algorithm is used for processing the gamma-ray spectrum. As previously mentioned, we can see that only Compton edges can be identified in the input gamma-ray spectrum. On the right, it is possible to see the result obtained after the deconvolution process in terms of incident energies. The main peaks given as input parameters of the simulation can be clearly identified (59.5 keV for ^{241}Am, 661.7 keV for ^{137}Cs, 1173.2 and 1332.5 keV for ^{60}Co). Moreover, as the H matrix integrates the efficiency of interest, the result after deconvolution direct gives an activity value (same intensity for instance for both peaks of ^{60}Co after deconvolution). However, it is important to mention that the incident energy obtained after the analysis corresponds to a value allowed by the grid of incident energies (for instance, considering the peak due to ^{137}Cs, the rebuilt incident energy is spread between 660 and 670 keV, due to the binning defined by the end user). **Figure 10** illustrates processing on the same simulated data but considering this time the database approach (deconvolution on a family of radionuclides and not on incident energies). It should be noted that results presented in **Figure 10** have been obtained using a different algorithm than ML-EM (nonparametric Bayesian methodology).

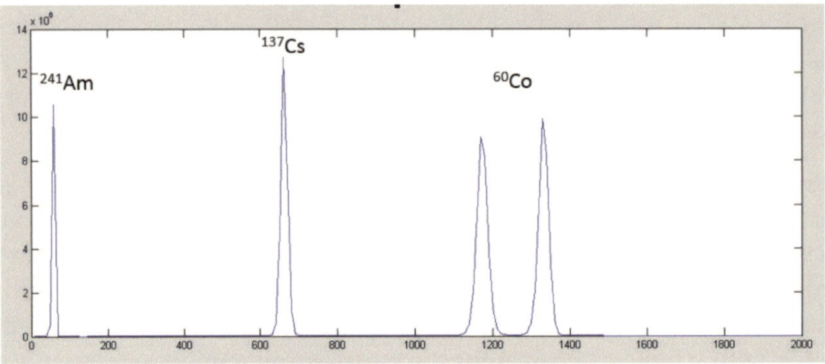

Figure 9. Incident energies of a gamma sources mixtures (^{241}Am, ^{137}Cs, and ^{60}Co) obtained after processing with an inverse problem approach [41].

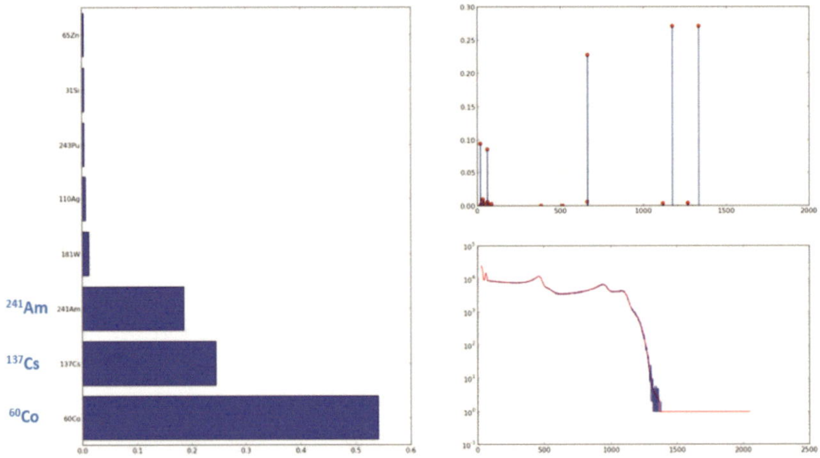

Figure 10. On the left: nature and associated proportion of radionuclides present in the gamma-ray spectrum. On the right: Gamma-rays associated with the detected radionuclides and comparison between simulated (blue curve) and rebuilt gamma-ray spectra (red curve).

3.5. Current status and future developments

Table 2 summarizes breakthroughs obtained in the field of metal loading.

http://dx.doi.org/10.5772/67134

Metal additive (wt%)	Organometallic compound	Matrix	Full energy peak absorption (keV)	Resolution (%)	γ light yield[a] (ph/MeV)	Typical sample size (cm³)	Ref.
Sn (7.8)	*p*-triphenylstyryltin	PVT	662	13	≈6000	n.g.	[10]
Sn (6)	Tributyltinmethacrylate	PSt	662	11.4	6700	6.4	[17]
Pb (n.g.)	n.g.	PSt	662	16	9000	n.g.	[19]
Bi (19.0)	Triphenylbismuth	PVK	662	6.8	7200	1	[21]
Bi (21.3)	Triphenylbismuth	PVK	1275	9	≈12,000	48	[22]
Bi (6)	Bismuth tripivalate	PVT	662	15	3300	103	[23]
Bi (5)	Triphenylbismuth	PSt	122	78	3900	155	[26]
Bi (8)	Acetyldimethacrylylbismuth	PSt	122	n.g.	2500	20.8	[24]
Gd (3.1)[b]	Gd_2O_3 nanocrystals	PVT	662	11.4	n.g.[c]	0.46	[28]
Yb (15)	YbF_3 nanoparticles	PVT	662	n.g.	8600	0.16	[31]
Hf (28.5)	HfO_2	PVT	662	8	≈10,000	0.16	[30]
Hf (10)	$Hf_xSi_{1-x}O_2$	PSt	67.4[d]	n.g.	4560	<1	[32]
Zr (22)	ZrO_2 nanoparticles	PSt	67.4[d]	n.g.	n.g.	<1	[33]

n.g. denotes to not given.
[a]When not explicitly given, evaluated from the gamma spectrum.
[b]Based on a calculated volume density.
[c]A beta light yield is given with ^{204}Tl excitation, showing 27,000 ph/MeV, relative to BC-400 plastic scintillator.
[d]X-rays excitation.

Table 2. Main improvements leading to gamma-ray spectrometry with plastic scintillators.

Gamma-ray counting or spectrometry enhancement with plastic scintillator loading is a pretty old strategy, but new key technologies allow now identifying gamma-ray spectra with various energies. This is extremely mandatory for the future generation radiation portal monitors used in homeland security. Most of the design is now performed with organometallic compounds mainly with tin, lead, and bismuth organometallics, but we have seen herein that other solutions may be of great value. Loading inorganics in plastic scintillators is obviously the cheapest method, but this method will be rapidly limited in terms of metal concentration and optical transparency. Metal-encapsulated nanoparticles would represent the most affordable and efficient in terms of gamma-ray spectrometry, but the described samples usually stand at the cm^3 state, so not big enough to find an application in radiation portal monitors.

Analysis of gamma-ray spectra using innovative methods is of great interest for identifying radionuclides, especially for homeland security applications where there is a real need for this type of features. Analysis methods based on a global processing of the gamma-ray spectrum are a performing way to identify radionuclides as soon as detectors with poor energy resolution are used. Several challenges can be identified as mid-term and long-term perspectives for developing and improving such algorithms. First of all, for homeland security applications, a second-order analysis is often required (for instance, if a moving object like a truck should be analyzed) and the processing time should be reduced as low as possible. Several teams developed identification solutions well adapted to address this challenge (see for instance Ref. [40]). Another issue strongly impacting this family of methods concerns the accuracy of the database used during the deconvolution step and its consistency with the gamma-ray spectrum to be measured. If the radioisotope is hidden by a specific shielding, the measured gamma-ray spectrum will be modified accordingly, and a bias will be introduced in comparison with the reference signature, potentially impacting the reconstruction step. Finally, coupling the algorithmic part with modified plastic scintillators, including for instance Bi loading, could improve the identification step because of the apparition of specific features in the spectrum, like full energy and escape peaks.

Author details

Matthieu Hamel* and Frédérick Carrel*

*Address all correspondence to: matthieu.hamel@cea.fr; frederick.carrel@cea.fr

CEA, LIST, Laboratoire Capteurs and Architectures Électroniques, CEA Saclay, Gif-sur-Yvette, France

References

[1] Schorr MG, Torney FL. Solid non-crystalline scintillation phosphors. Physical Reviews 1950;**80**:474. doi:10.1103/PhysRev.80.474

[2] Förster T. Zwischenmolekulare Energiewanderung und Fluoreszenz. Annalen der Physik 1948;**437**(1–2):55–75. doi:10.1002/andp.19484370105

[3] Bertrand GHV, Hamel M, Sguerra F. Current status on plastic scintillators modifications. Chemistry – A European Journal 2014;**20**(48):15660–15685. doi:10.1002/chem.201404093

[4] Evans RD. The Atomic Nucleus. McGraw-Hill Company; 1955. p. 712. doi:10.1002/aic.690020327

[5] Pichat L, Pesteil P, Clément J. Solides fluorescents non cristallins pour mesure de radioactivité. Journal de Chimie Physique et de Physico-Chimie Biologique 1953;**50**:26–41.

[6] Cho ZH, Tsai CM, Eriksson LA. Tin and lead loaded plastic scintillators for low energy gamma-ray detection with particular application to high rate detection. IEEE Transactions on Nuclear Science 1975;**NS-22**(1):72–80. doi:10.1109/TNS.1975.4327618

[7] Zhao YS, Yu Z, Douraghy A, Chatziioannou AF, Mo Y, Pei Q. A facile route to bulk high-Z polymer composites for Gamma-ray scintillation. Chemical Communications 2008:6008–6010. doi:10.1039/b813571a

[8] Britvich GI, Vasil'chenko VG, Lapshin VG, Solov'ev AS. New heavy plastic scintillators. Instruments and Experimental Techniques 2000;**43**(1):36–39. doi:10.1007/BF02758995

[9] Basile LJ. Characteristics of plastic scintillators. The Journal of Chemical Physics 1957;**27**:801–806. doi:10.1063/1.1743832

[10] Sandler SR, Tsou KC. Quenching of the scintillation process in plastics by organometallics. The Journal of Physical Chemistry 1964;**68**:300–304. doi:10.1021/j100784a015

[11] Hyman Jr. M, Ryan JJ. Heavy elements in plastic scintillators. IRE Transactions on Nuclear Science. 1958;**5**(3):87–90. doi:10.1109/TNS2.1958.4315631

[12] Baroni EE, Kilin SF, Lebsadze TN, Rozman IM, Shoniya VM. Addition of hetero-organic compounds to polystyrene. Soviet Atomic Energy 1964;**17**(6):1261–1264. doi:10.1007/BF01122774

[13] Sandler SR, Tsou KC. Evaluation of organometallics in plastic scintillators towards gamma-radiation. The International Journal of Applied Radiation and Isotopes 1964;**15**(7):419–426. doi:10.1016/0020-708X(64)90140-1

[14] Dannin J, Sandler SR, Baum B. The use of organometallic compounds in plastic scintillators for the detection and resolution of Gamma-rays. The International Journal of Applied Radiation and Isotopes 1965;**16**(10):589–597. doi:10.1016/0020-708X(65)90095-5

[15] Sandler SR, Dannin J, Tsou KC. Copolymerization of *p*-Triphenyltinstyrene and *p*-Triphenylleadstyrene with Styrene or Vinyltoluene. Journal of Polymer Science I Part A, Polymer Chemistry 1965;**3**:3199–3207. doi:10.1002/pol.1965.100030914

[16] Feng PL, Mengesha W, Anstey MR, Cordaro JG. Distance dependent quenching and gamma-ray spectroscopy in tin-loaded polystyrene scintillators. IEEE Transactions on Nuclear Science 2016;**63**(1):407–415. doi:10.1109/TNS.2015.2510960

[17] Hamel M, Turk G, Rousseau A, Darbon S, Reverdin C, Normand S. Preparation and characterization of highly lead-loaded red plastic scintillators under low energy

X-rays. Nuclear Instruments and Methods in Physics Research Section A: Accelerators, Spectrometers, Detectors and Associated Equipment 2011;**660**(1):57–63. doi:10.1016/j.nima.2011.08.062

[18] van Loef E, Markosyan G, Shirwadkar U, McClish M, Shah K. Gamma-ray spectroscopy and pulse shape discrimination with a plastic scintillator. Nuclear Instruments and Methods in Physics Research Section A: Accelerators, Spectrometers, Detectors and Associated Equipment 2015;**788**:71–72. doi:10.1016/j.nima.2015.03.077

[19] Bertrand GHV, Hamel M, Normand S, Sguerra F. Pulse shape discrimination between (fast or thermal) neutrons and Gamma-rays with plastic scintillators: State of the art. Nuclear Instruments and Methods in Physics Research Section A: Accelerators, Spectrometers, Detectors and Associated Equipment 2015;**776**:114–128. doi:10.1016/j.nima.2014.12.024

[20] Rupert BL, Cherepy NJ, Sturm BW, Sanner RD, Payne SA. Bismuth-loaded plastic scintillators for gamma-ray spectroscopy. Europhysics Letters 2012;**97**:22002. doi:10.1209/0295-5075/97/22002

[21] Cherepy NJ, Sanner RD, Tillotson TM, Payne SA, Beck PR, Hunter S, Ahle L, Thelin PA. Bismuth-loaded plastic scintillators for gamma spectroscopy and neutron active interrogation. IEEE Nuclear Science Symposium and Medical Imaging Conference Record (NSS/MIC) 2012:1972–1973. doi:10.1109/NSSMIC.2012.6551455

[22] Cherepy NJ, Sanner RD, Beck PR, Swanberg EL, Tillotson TM, Payne SA, Hurlbut CR. Bismuth- and lithium-loaded plastic scintillators for gamma and neutron detection. Nuclear Instruments and Methods in Physics Research Section A: Accelerators, Spectrometers, Detectors and Associated Equipment 2015;**778**:126–132. doi:10.1016/j.nima.2015.01.008

[23] Bertrand GHV, Sguerra F, Dehé-Pittance C, Carrel F, Coulon R, Normand S, Barat É, Dautremer T, Montagu T, Hamel M. Influence of bismuth loading in polystyrene-based plastic scintillators for low energy gamma spectroscopy. Journal of Materials Chemistry C 2014;**2**:7304–7312. doi:10.1039/c4tc00815d

[24] Koshimizu M, Bertrand GHV, Hamel M, Kishimoto S, Haruki R, Nishikido F, Yanagida T, Fujimoto Y, Asai K. X-ray Detection Capabilities of bismuth-doped Plastic Scintillators. Japanese Journal of Applied Physics 2015;**54**:102202. doi:10.7567/JJAP.54.102202

[25] Bertrand GHV, Dumazert J, Sguerra F, Coulon R, Corre G, Hamel M. Understanding behavior of different metals in loaded scintillators: discrepancy between gadolinium and bismuth. Journal of Materials Chemistry C 2015;**3**:6006–6011. doi:10.1039/c5tc00387c

[26] Carturan SM, Marchi T, Maggioni G, Gramegna F, Degerlier M, Cinausero M, Palma MD, Quaranta A. Thermal neutron detection by entrapping ^{6}LiF nanocrystals in siloxane scintillators. Journal of Physics: Conference Series 2015;**620**:012010. doi:10.1088/1742-6596/620/1/012010

[27] Cai W, Chen Q, Cherepy N, Dooraghi A, Kishpaugh D, Chatziioannou A, Payne S, Xiang W, Pei Q. Synthesis of bulk-size transparent gadolinium oxide–polymer nanocomposites for Gamma-ray spectroscopy. Journal of Materials Chemistry C 2013;**1**:1970–1976. doi:10.1039/c2tc00245k

[28] Chen Q, Cai W, Hajagos T, Kishpaugh D, Liu C, Cherepy N, Dooraghi A, Chatziioannou A, Payne S, Pei Q. Bulk-Size Transparent Gadolinium Oxide-Polymer Nanocomposites for Gamma-ray Scintillation. Nanotechnology 2014: Graphene, CNTs, Particles, Films & Composites, Technical Proceedings of the NSTI Nanotechnology Conference and Expo 2014:**1**:295–296. ISBN: 978-1-4822-5826-4

[29] Liu C, Hajagos TJ, Kishpaugh D, Jin Y, Hu W, Chen Q, Pei Q. Facile single-precursor synthesis and surface modification of hafnium oxide nanoparticles for nanocomposite γ-ray scintillators. Advanced Functional Materials 2015;**25**(29):4607–4616. doi:10.1002/adfm.201501439

[30] Jin Y, Kishpaugh D, Liu C, Hajagos TJ, Chen Q, Li L, Chen Y, Pei Q. Partial ligand exchange as a critical approach to synthesizing transparent ytterbium fluoride-polymer nanocomposite monoliths for Gamma-ray scintillation. Journal of Materials Chemistry C 2016;**4**:3654–3660. doi:10.1039/C6TC00447D

[31] Sun Y, Koshimizu M, Yahaba N, Nishikido F, Kishimoto S, Haruki R, Asai K. High-energy X-ray detection by hafnium-doped organic-inorganic hybrid scintillators prepared by sol-gel method. Applied Physics Letters 2014;**104**:174104. doi:10.1063/1.4875025

[32] Araya Y, Koshimizu M, Haruki R, Nishikido F, Kishimoto S, Asai K. Enhanced detection efficiency of plastic scintillators upon incorporation of zirconia nanoparticles. Sensors and Materials 2015;**27**(3):255–261. doi:10.18494/SAM.2015.1063

[33] Demkiv TM, Halyatkin OO, Vistovskyy VV, Gektin AV, Voloshinovskii AS. Luminescent and kinetic properties of the polystyrene composites based on BaF_2 nanoparticles. Nuclear Instruments and Methods in Physics Research Section A: Accelerators, Spectrometers, Detectors and Associated Equipment 2016;**810**:1–5. doi:10.1016/j.nima.2015.11.130

[34] Java-based nuclear information software (JANIS). Available from: http://www.nea.fr/janis [accessed 2016-07-26].

[35] Briesmeister JF. MCNP™ – A General Monte Carlo N-Particle Transport Code – Version 4C, Los Alamos National Laboratory report LA-13709-M, 2000.

[36] Phillips DL. A technique for the numerical solution of certain integral equations of the first kind. Journal of the ACM 1962;**9**(1):84–97. doi:10.1145/321105.321114

[37] Jaynes ET. How does the brain do plausible reasoning? In Erickson GJ, Smith CR, editors. Maximum-Entropy and Bayesian Methods in Science and Engineering. Foundations. Kluwer Academic Publishers; 1988. Vol 1. pp. 1–24. doi:10.1007/978-94-009-3049-0_1

[38] Shepp LA, Vardi Y. Maximum likelihood reconstruction for emission tomography. IEEE Transactions on Medical Imaging 1982;**1**(2):113–122. doi:10.1109/TMI.1982.4307558

[39] Burt C. Plastic Scintillation Spectrometry. PhD dissertation. Southampton University; 2009. Available from: http://eprints.soton.ac.uk/72351/1.hasCoversheetVersion/Thesis.pdf [last accessed 2017-01-05].

[40] Corre G, Boudergui K, Sannié G, Kondrasovs V. A Generic Isotope Identification Approach for nuclear instrumentation, In: IEEE proceedings of Advancements in Nuclear Instrumentation Measurement Methods and their Applications; 20–24 April 2015; Lisbon. IEEE; 2016. doi:10.1109/ANIMMA.2015.7465628

[41] Hamel M, Dehé-Pittance C, Coulon R, Carrel F, Pillot P, Barat É, Dautremer T, Montagu T, Normand S. Gammastic: towards a pseudo-gamma spectrometry in plastic scintillators. In: IEEE proceedings of Advancements in Nuclear Instrumentation Measurement Methods and their Applications; 23–27 June 2013; Marseille. IEEE; 2014. doi:10.1109/ANIMMA.2013.6727889

[42] EJ-256 data sheet. Available from: http://www.eljentechnology.com/index.php/products/plastic-scintillators/ej-256 [last accessed 2016-07-26].

Section 2

Gamma Rays In Space

Chapter 4

Gamma Rays from Space

Carlos Navia and Marcel Nogueira de Oliveira

Additional information is available at the end of the chapter

http://dx.doi.org/10.5772/67176

Abstract

An overview of gamma rays from space is presented. We highlight the most powerful astrophysical explosions, known as gamma-ray bursts. The main features observed in detectors onboard satellites are indicated. In addition, we also highlight a chronological description of the efforts made to observe their high energy counterpart at ground level. Some candidates of the GeV counterpart of gamma-ray bursts, observed by Tupi telescopes, are also presented.

Keywords: gamma-ray astrophysics, cosmic rays, particle detectors

1. Introduction

Gamma rays are the most energetic form of electromagnetic radiation, with a very short wavelength of less than 0.1 nm. Gamma radiation is one of the three types of natural radioactivity discovered by Becquerel in 1896. Gamma rays were first observed in 1900 by the French chemist Paul Villard when he was investigating radiation from radium [1].

They are emitted by a nucleus in an excited state. The emission of gamma rays does not alter the number of protons or neutrons in the nucleus. Gamma emission frequently follows beta decay, alpha decay, and other nuclear decay processes.

On the other hand, cosmic-ray particles (mostly protons) that arrive at the top of the Earth's atmosphere are termed primaries; their collisions with atmospheric nuclei give rise to secondaries. These secondary particles are constituted by pions (subatomic particles); the dominant decay of a neutral pion is the electromagnetic decay in two photons (two gamma rays). This process is the origin of the highest energy gamma rays.

Nowadays, the artificial production of pions (and consequently gamma rays) in N-N collisions is produced copiously. The artificial production of pions started in 1948, with Lattes and Gadner using the 184-inch synchrocyclotron at Lawrence Berkeley National Laboratory (California) which accelerated protons to 350 MeV [2, 3].

As gamma rays coming from space are absorbed by the Earth's atmosphere, the first detection of gamma rays coming from space was through satellites. Indeed, the Explorer satellite in 1961 confirmed the existence of gamma rays in space [4]. Gamma rays coming from space have frequencies greater than about 10^{18} Hz; they occupy the same region of the electromagnetic spectrum as hard X-rays or above them. The only difference between them is their source: X-rays are produced by accelerating electrons, whereas gamma rays are produced by atomic nuclei decays and/or nuclear collisions.

There is a large variety of gamma-ray sources from space: the most important is the Sun's transient activity, such as solar flares and coronal mass ejection (CME). This means that there is also a "gamma-ray background" due to other stars in our galaxy and from the stars of other galaxies, as well as that expected from the interaction of cosmic rays (very energetically charged particles in space) with interstellar gas. There are also sources of gamma rays with continuous emission, or long-lived gamma-ray emission from compact sources. This emission is in the high-energy region from GeV to TeV. The Fermi Large Area Telescope (LAT) produced an inventory of 1873 objects shining in gamma-ray light [5].

The most intense are the galactic objects, such as the Crab nebula (Taurus constellation) and the W44 nebula (Aquila constellation), both two supernova remnants. There are also bright extragalactic objects, such as the objects called active galactic nuclei (AGN); the most bright are the so-called Markarian 501 and 421 AGNs. These extragalactic objects were discovered by a ground-based gamma-ray telescope, the "air Cherenkov" Wipple telescope [6], confirmed by telescopes mounted on satellites.

Gamma-ray bursts (GRBs) are the most energetic explosions in the Universe: they are bright flashes of gamma rays for a short period of time, in most cases less than 100 s, with a photon's flux of around 0.1–100 ph/cm^2/s/keV at Earth. There is some evidence that indicates that a GRB of long duration (i.e., above 2 s). They are associated with exploding massive stars called as Hypernovas. The signature of these explosions is two opposing beams of gamma rays and if any one of these beams is in the Earth's direction, we will see the gamma-ray burst.

In addition, sometimes, several narrow lines are detected; this radiation comes from the material around the explosion that was excited by the blast and that permits the detection of its host galaxy. However, sometimes, no narrow gamma-ray lines are detected, especially in GRBs of short duration (less than 2 s), meaning that some of these bursts come from different progenitors, such as merging compact objects such as black holes or neutron stars, or perhaps from a massive star explosion, but without the formation of a supernova.

http://dx.doi.org/10.5772/67176

In most cases, the gamma rays from GRBs, observed by space crafts, have energies in the keV region; a fraction has photons in the MeV region, called a high-energy counterpart. In some cases, photons can reach energies up to several dozen of GeV. In this case, the photons have sufficient energy to produce other particles when they reach the upper atmosphere, i.e., particles in the air, forming showers via cascading processes. This suggests the possibility of detecting gammas from the ground level, at least those that are more energetic and relatively long-term.

Despite great efforts for the systematic observation of GRBs at ground level, the results have been negative. However, a few events have been detected with a high significance, by the Tupi experiment in Brazil. They are considered as good candidates. Here, in addition to giving an overview of the GRBs, we will highlight these remarks (i.e., the Tupi events) and their implications on GRB physics.

2. General features of GRBs

The first scientific paper on GRBs was published in 1973 [7]. This paper reported the observation of 73 GRBs, starting on 2 July 1967, based on data from US satellites (Vela project) designed to monitor Russian nuclear weapon tests in space.

However, this first work did not answer several fundamental questions on the origin and nature of GRBs; for instance, are they from the solar system, Galactic halo, or are they extragalactic objects?

The mystery persisted until 1998, when the first spatial distribution of GRBs came from BATSE gamma detector and the more energetic (MeV to GeV) by EGRET, both onboard the Compton Gamma-Ray Observatory (CGRO). It was one of the best space observatories for detecting gamma rays from 20 keV to 30 GeV in Earth's orbit from 1991–2000. The BATSE instrument detected gamma-ray burst, at a rate of one per day, with a total of approximately 2700 detections. An isotropic sky distribution of GRBs was reported by BATSE. This means an extragalactic origin for the GRBs.

These observations were confirmed by new observations made by the new generations of GRBs detectors onboard satellites, such as the Swift, which is a multiwavelength GRB detector [8] with a wide field of view being able detect more than 100 GRBs per year. Swift has the Burst Alert Telescope (BAT) which covers the 15–150 keV energy band, the X-ray telescope (XRT), and an ultraviolet and optical telescope (UVOT) to detect X-ray and UV optical afterglows.

Swift has several online (free access) catalogs of GRBs. **Figure 1** shows the location in the sky (equatorial coordinate system) of 1188 Swift GRBs, detected from 17 December 2004 to 25 May 2016.

Thus, Swift confirmed the isotropic distribution of GRBs.

However, only the discovery of the first X-ray afterglows in 1998 by the BeppoSax satellite [9] allowed the accurate positions and the identification of the γ-ray afterglow with 'normal' galaxies to be obtained. It was also possible to obtain the redshifts and consequently their distances, confirming that GRBs have a cosmological origin.

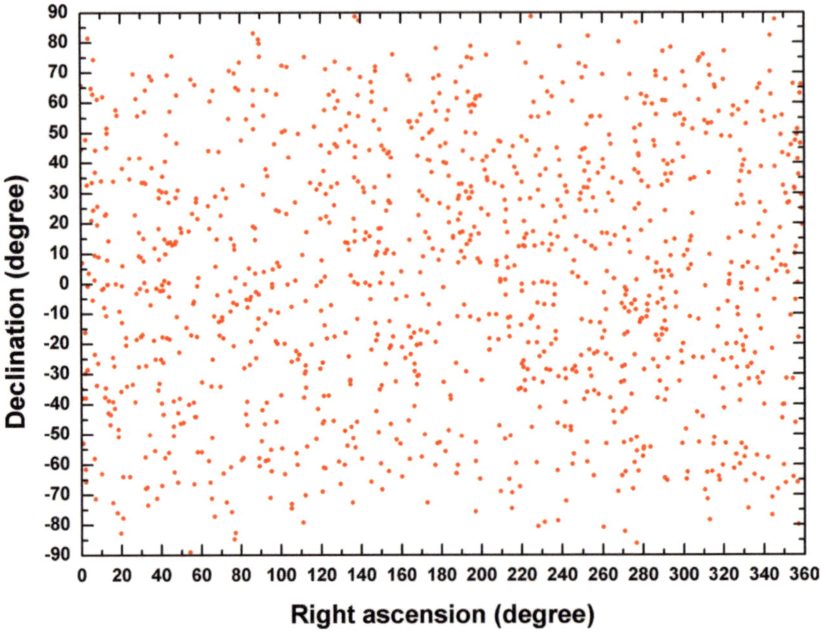

Figure 1. Location in the sky (equatorial coordinate system) of 1188 Swift GRBs, detected from 17 December 2004 to 25 May 2016.

It was also the BATSE gamma detector onboard the CGRO that showed that there are two different types of GRBs; that is, GRBs are separated into two classes: long-duration bursts, normally from 2 to 500 s, with an average duration of about 30 s; and short-duration bursts, ranging from a few ms to 2 s, with an average duration of only 0.3 s, following a bimodal distribution. They are not small and large versions of the same phenomenon; the two types of bursts have completely different sources. The origin of long-duration bursts is linked to supernovae explosions of massive stars (with masses 100 times greater than the solar mass), called hypernovae. Short-duration bursts are linked to the merging of two objects, such as two neutron stars, or a neutron star and a black hole, or two black holes.

This bimodal distribution was also confirmed by Swift as shown in **Figure 2** (top panel). The figure includes also a scatter plot (bottom panel), a correlation between the T90 duration of the GRB and the integrated time fluence. T90 is defined as the time interval over which 90% of the total background-subtracted counts are observed, with the interval starting when 5% of the total counts have been observed. In addition, the integrated time fluence is the energy deposited on the detector per unit area, during the T90 duration of the GRBs. If the distance of the source of the GRB is determined, this last quantity allows estimate the energy released during the explosion. The energy output of GRBs on some cases, if spherically radiated is above 1054 erg. This exceeds any reasonable source during such a short timescale, so the radiation is likely highly beamed.

http://dx.doi.org/10.5772/67176

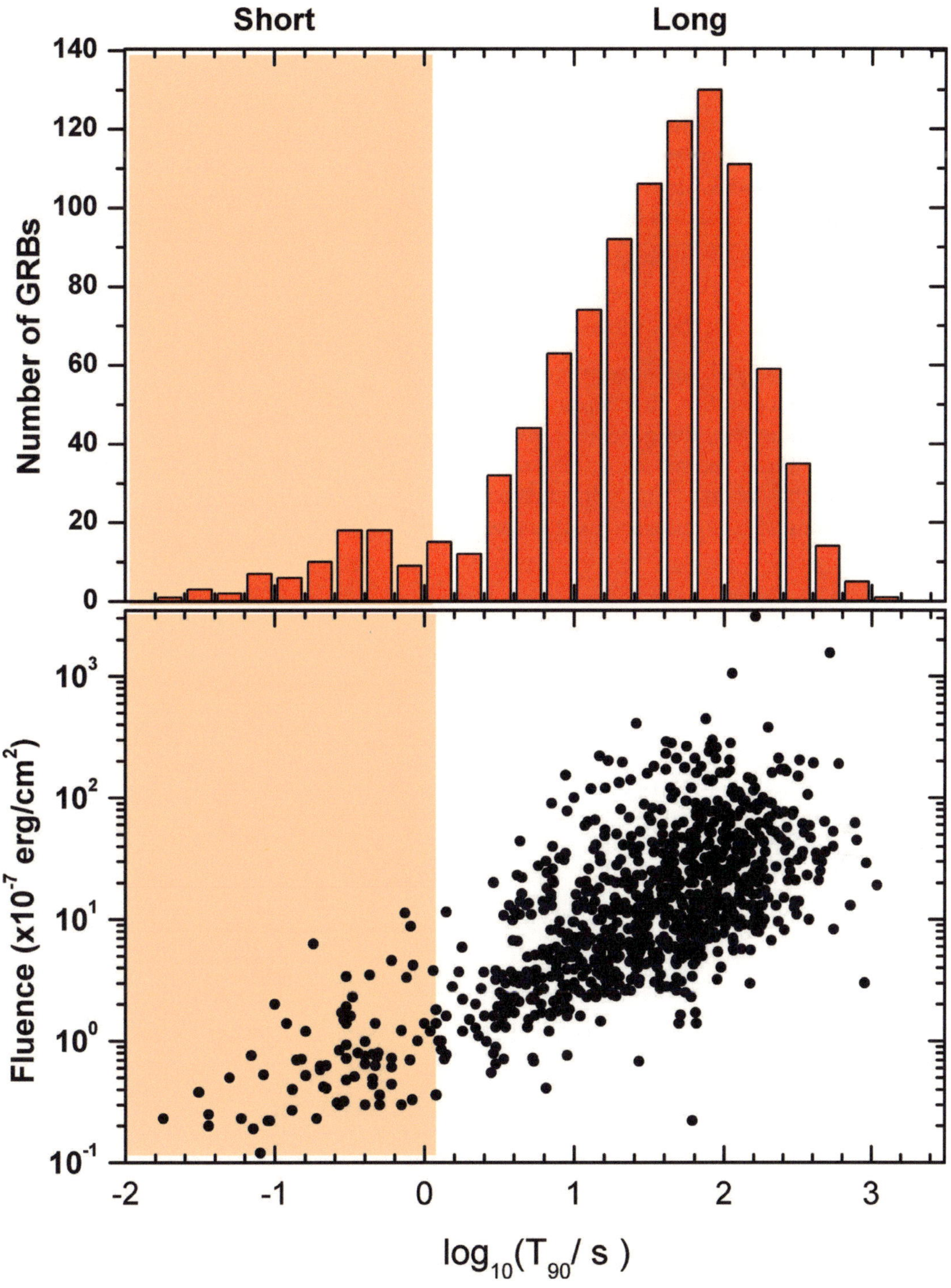

Figure 2. Top panel: bimodal classification of Swift GRBs, as short and long. Bottom panel: correlation between the T90 duration of the GRBs and the integrated time fluence, on the basis of 1188 Swift GRBs, detected from 17 December 2004 to 25 May 2016.

3. Emission mechanisms

Despite remarkable progress in the past few years by theory and breakthroughs of observations, our understanding of these fascinating cosmic events is still very incomplete. Many aspects remain uncertain and demand further exploration. For instance, the detailed physics of the central engine is poorly understood. Even so, some GRB models can reproduce the main features of the observed bursts, irrespective of the detailed physics of the central engine.

The relativistic fireball GRB model was introduced in the 1990s [10, 11] on the basis of earlier works [12–14]. Although the model does not explain the central engine of a GRB, it has been successful in explaining the various features of GRBs, such as the origin of their afterglows.

According to this model, GRBs are produced far from the source (10^{11}–10^{12} m), through the interactions between the outflow (fireball) and the surrounding medium (internal shocks). Following **Figure 3** that illustrates the generation mechanism of the GRBs, we can see that the X-rays afterglow result from the subsequent interaction of the outflow (fireball) with the surrounding medium (external shocks). The outflow fireball ends up losing its kinetic energy through successive interactions with the external medium, resulting in UV, visible, and radio afterglows.

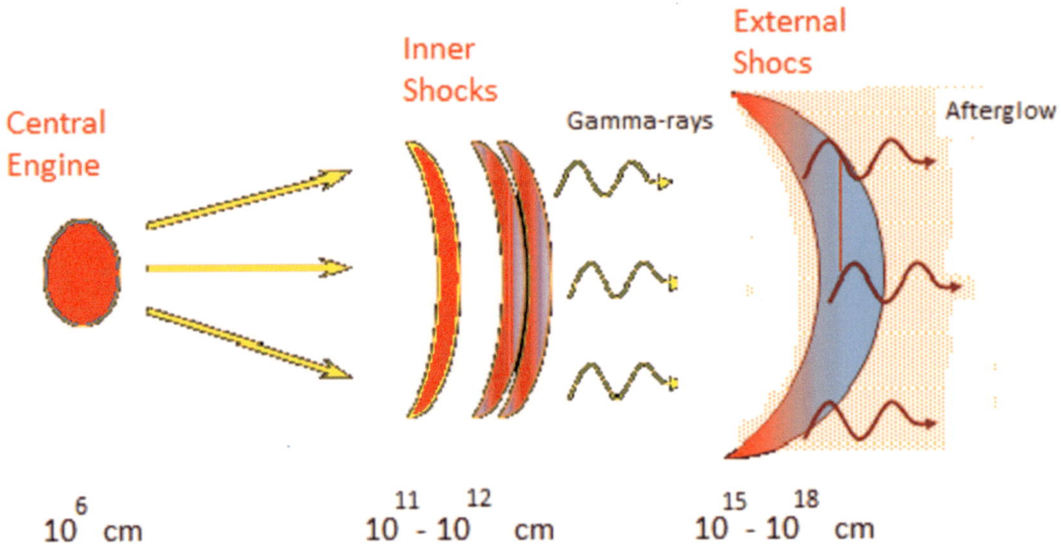

Figure 3. Fireball model scheme of the generation mechanism of the GRBs. The GRBs are generated far from the source, through the interactions between the outflow (fireball) and the surrounding medium (internal shocks). The X-ray, UV, visible, and radio afterglow result from the subsequent interaction of the outflow (fireball) with the surrounding medium (external shocks).

The time variability of the gamma rays is 10^{-3} s, meaning a size of the emitting region of around 10^5 m; that is, a relativistic fireball, with a Lorentz factor above 1000 ($\Gamma > 1000$). The typical energy emitted in a collimated beaming flux is around 10^{45} J. A high Lorentz factor Γ also allows a relativistic collimated jet, with an aperture of $\theta \sim 1/\Gamma$. As a consequence of this behavior, the relative angle at which photons collide decreases and leads to an increase in the pair production threshold.

However, the simple relativistic fireball model produces a modified blackbody spectrum. This mechanism converts energy into thermal energy efficiently, thus it is necessary to reconvert kinetic energy into nonthermal emission, which happens when the fireball becomes optically thin. Thus, the reconverted kinetic energy into random energy must be via shocks, after the flow becomes optically thin (mainly synchrotron radiation).

In short, the fireball model can reproduce the main features of the observed bursts, irrespective of the detailed physics of the central engine.

Many GRB afterglow models [15–17] predict the production of photons in the GeV to TeV energy range, and GeV emission has indeed been detected by previous [18] and current-generation (Fermi LAT) space-based ray detectors [19].

The several GRBs observed by EGRET and Fermi LAT in the GeV energy region, as well as the ground-based observations of GRB candidates by Tupi suggest that the energy spectrum extend beyond GeV energies. In addition, there are some mechanism to explain this extension to very high energies, such as the synchrotron selfCompton model [20, 21], which provides a natural explanation for the optical and gamma-ray correlation seen in some GRBs. It has also been shown that a relatively strong second-order inverse Compton (IC) component of the GRB spectrum should peak in the 10s GeV energy region [22].

Another proposed model is the cannonball (CB). The cannonball model is inspired by observations of quasars and microquasars [23–25] and assumes that a supermassive star, when entering the final phase of its life, undergoes gravitational collapse, becoming a supernova (SN). In this internship, an accretion toroidal disk is developed around a compact object that is newly formed. The matter is then ejected as a CB in bipolar jets of plasma droplets (plasmoids) that are highly relativistic.

These jets collide with the photons inside the star through an inverse Compton scattering process, producing gamma rays. In this model, each pulse of GRB produced during the collapsing star corresponds to a CB.

The range of emission of these gamma rays is related to the layers (shields) of the star's interior, where these missiles collide. The CBs are individually ejected and the light curve observed in GRB depends on the local emission properties. As with the fireball model, the CB model can also describe the afterglows, such as X-ray, UV, and radio flares. The CB model also includes a description of other phenomena, such as the acceleration of cosmic rays (CRs) in a successful way [26].

The observed energy spectra of gamma-ray bursts reveal a diverse phenomenology. The spacecraft observed gamma-rays up to 33 GeV [27]. While some energy spectra have been fitted by a simple expression over many decades [28], others require a few separate components to explain high-energy emission [27]. In most cases (at low energies), the GRB spectrum is well described by a phenomenological "band function" in a "Comptonized model" using a power law with an exponential cutoff:

$$N(E) = K E^{-\alpha} e^{-E/E_0} \tag{1}$$

where α is the power-law exponent and E_0 is the cutoff energy. At high energies, the spectrum is described well by a power-law function with a steeper slope:

$$N(E) = A_\gamma E^{-\beta} \tag{2}$$

where $\alpha > \beta$, and the spectral parameters α, β, and E_0 vary from burst to burst. For instance, a "blast wave model," usually considered for GRB sources, is quite sensitive to the relationship between these two power-law indices.

4. Ground level observations

We present in this section a brief description of the various efforts for detecting at ground level, the GeV to TeV counterpart of the GRBs.

4.1. Milagrito

Milagrito was a detector (air shower detector, the predecessor of the Milagro detector) sensitive to very high-energy gamma rays, which monitored the northern sky with a large field of view and a high duty cycle, located near the Los Alamos Laboratory (New Mexico, USA). This instrument was well suited to perform a search for TeV gamma-ray bursts (GRBs). From February 1997 through May 1998, BATSE (Transient Satellite experiment) aboard the Compton Gamma-Ray Observatory detected 54 GRBs within the field of view of Milagrito.

The Milagrito results were negative; that is, no significant correlations were detected from the other bursts, with the exception of a single event, which was reported as evidence of a marginal emission at TeV energies from GRB 970417a. The event had a chance probability of 2.8×10^{-5} of being background fluctuation. The probability of observing an excess at least this large from any of the 54 bursts is 1.5×10^{-3} [29].

4.2. Milagro

Milagro was a wide-field (2 sr) high-duty cycle (>90%) ground-based water Cherenkov detector (60 m wide × 80 m long × 8 m deep) located at 2630 m above sea level in the Jemez mountains, New Mexico. Milagro had 723 PMTs divided into two layers under water. It triggers mainly on extensive air showers (EAS) in the energy range of 100 GeV to 100 TeV.

Milagro operated from January 2000 to May 2008. The gamma-ray coordinates network (GCN) system incorporated the distribution of positions of GRBs and transients detected by the MILAGRO instrument. However, none of these events was confirmed as true GRBs. Even so, Milagro succeeded in the detection of gamma rays in the TeV energy region, such as TeV gamma rays from the galactic plane [30], and the discovery of TeV gamma ray emission from the Cygnus region of the Galaxy [31]. Perhaps, its high energy threshold (above 100 GeV) set for gamma rays did not allow the detection of GRBs; thus only the upper limits were reported [32].

4.3. ARGO

The Astrophysical Radiation with Ground-based Observatory at Yangbajing, China (Tibet—4300 m a.s.l.), under the auspices of the ARGO-YBJ experiment, is through an air shower detector. The ARGO-YBJ detector has a large active surface of around 6700 m^2 of Resistive Plate Chambers, a wide field of view ~2 sr, and a high duty cycle (>86%). The ARGO-YBJ experiment is a collaboration of Italian and Chinese institutions [33].

The ARGO-YBJ performed a search for gamma-ray bursts (GRB) emission in the energy range 1–100 GeV in coincidence with satellite detection. From 17 December 2004 to 7 February 2013, a total of 206 GRBs occurring within the ARGO-YBJ field of view (zenith angle $\theta = 45°$) were analyzed, no significant excess was found, and only the corresponding fluence upper limits in the 1–100 GeV energy region were derived, with values as low as 10^{-5} erg/cm^2.

4.4. HAWC

The HAWC gamma-ray observatory is a wide field of view, continuously operating, TeV gamma-ray telescope that explores the origin and solar modulation of cosmic rays and searches for new TeV physics. HAWC is located at a high altitude of 4100 m above sea level in Mexico (Sierra Negra) and is a collaboration of 15 US and 12 Mexican institutions.

HAWC consists of an array of 300 water Cherenkov detectors and is expected to be more than one order of magnitude, which is more sensitive than its predecessor, Milagro. HAWC monitors the northern sky and makes coincident observations with other wide field of view observatories. The HAWC experiment is particularly suitable to detect short and unexpected events like GRBs. However, thus far, no excess candidate events, nor GRB counterparts, have been reported.

Many other experiments have searched for the GeV-TeV counterpart of GRBs. No conclusive detection such as INCA [34], Tibet AS [35], HEGRA AIROBICC [36], GRAND [37], LAGO [38], and the Cherenkov detector MAGIC have given a very low upper limit between 85 and 1000 GeV [39]. In short, no significant correlations among events of these experiments and Satellite GRBs observations have been related.

5. Ground level observation of gamma-ray bursts from space

The Earth's magnetic field effects on the development of a particle shower in the atmosphere spread the collecting particles and therefore decrease the sensitivity of the detector. This deflection is caused by the component of the Earth's magnetic field perpendicular to the particle trajectory. Thus, if the source of gamma rays is close to the vertical direction, the places with the smaller horizontal magnetic component will be the best places, to ground level detection of GRBs.

We point out that the location of the Tupi detector is within the South Atlantic Anomaly (SAA). It is the region characterized by anomalously weak geomagnetic field strength. It is the lowest magnetic field of the world. The SAA central region is located on 26° S, 53° W, over the South Atlantic Ocean on Brazil, close to the Tupi detector location. The horizontal geomagnetic field component is only 18.13 mT, that is, almost half than the horizontal geomagnetic field component of other locations, where a search for the detection of GRBs at ground level was performed. For details, please see Section 7 of the reference [40].

Since August 2013, the Tupi experiment has operated an extended array of five muon telescopes [40], located in Niteroi, Rio de Janeiro, Brazil, (22.9° S, 43.0° W).

The first has a vertical orientation. The other four have orientations to the north, south, east, and west; each telescope is inclined 45° relative to the vertical.

Each telescope was constructed on the basis of two detectors (plastic scintillators 50 × 50 × 3 cm) separated by a distance of 3 m. The coincident signals in the upper and lower Tupi detectors are registered at a rate of 1 Hz.

There are two flagstones of concrete (150 g cm^{-2}) above telescopes and only particles (muons) with energies above 0.1–0.2 GeV can penetrate the two flagstones. This defines the energy threshold of the telescopes. Each Tupi telescope has an effective field of view of 0.37 sr. For the vertical telescope, this corresponds to an aperture (zenith angle) of 20°.

We present ground level observations in the GeV energy range of possible counterparts associated with the gamma-ray bursts observed by spacecraft detectors [41], such as the MAXI onboard the ISS and the BAT onboard Swift [42].

In the period from 8 September 2013 to 10 August 2014, 34 GRBs observed by satellites in the keV energy region were within the field of view of the five Tupi telescopes. The majority of the events was compatible with the Tupi background fluctuations. The exceptions were the two events that are described below.

In addition, of the 34 GRBs with energies above 100 MeV observed by the Femi LAT detector up until 10 August 2014 [43], only one GRB131018B had their trigger coordinates within the field of view of one Tupi telescope. However, no signal was found. **Figure 4** shows the location in equatorial coordinates of the four Tupi telescopes and includes the trigger coordinates of the Fermi LAT GRBs until 10 August 2014.

5.1. Association with the MAXI gamma-ray burst

On 15 October 2013 at 21:55:44 UT, a peak (muon excess) with a significance of 5 s at the 68% confidence level was found in the 24 h raw data (counting rate 1 Hz) of the vertical Tupi telescope.

It was possible to recognize this peak in the time profile of the muon counting rate, just by the naked eye, as shown in **Figure 5**. The peak was found at To + 25.7 s, where To = 21:55:19 UT is the occurrence of the MAXI trigger [44]. The trigger coordinates of MAXI trigger were within the field of view of the vertical Tupi telescope.

In addition, a second narrow peak with a significance of 4 sigma (1 s binning) can be observed at To + 297.2 s. This peak can be seen also at the 3 and 5 s binning counting rates, as shown in **Figure 5**. This behavior strongly suggests that this peak is a true signal.

In order to see the background fluctuations more accurately, we examined the time profiles up to 30 min before and after the trigger time. A confidence analysis was made for a 1 h interval around the MAXI trigger time, as shown in **Figure 6**. In the absence of a signal, the background fluctuation of the counting rate follows a Gaussian distribution. Thus, the trials out of the Gaussian curve are considered as a signal's signature.

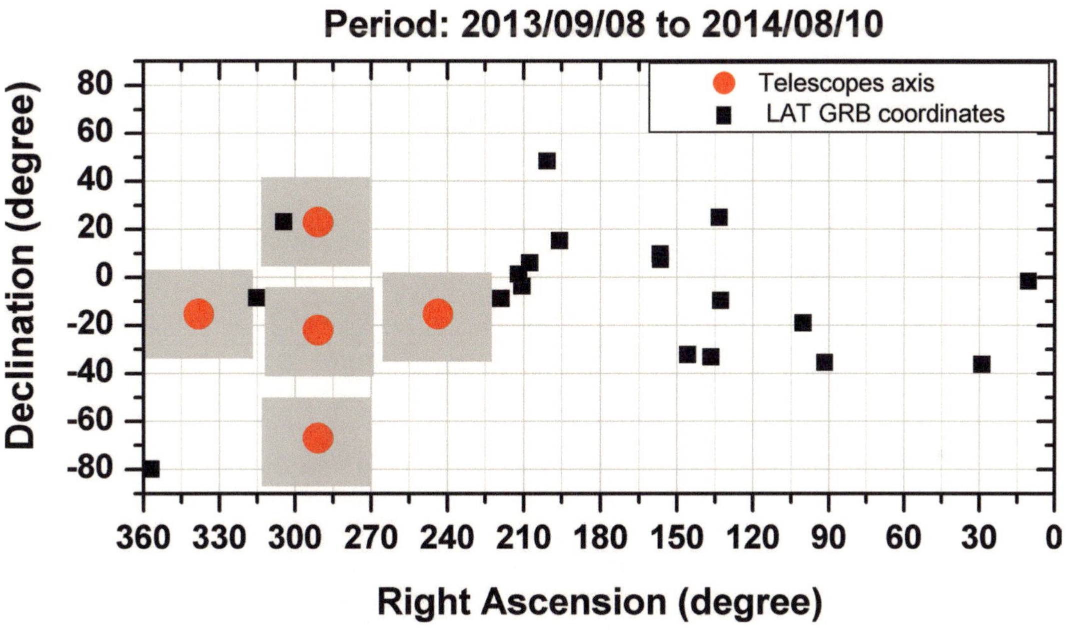

Figure 4. Equatorial coordinates showing the positions of the five Tupi telescope axes, as well as the Fermi LAT GRBs (>100 MeV) in the period until 10 August 2014. The squares with circles represent the FOV of the Tupi telescopes, which were within the FOV of the North Tupi telescope.

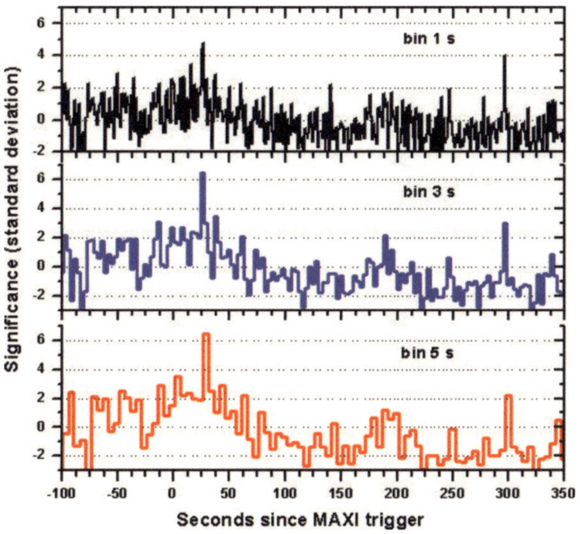

Figure 5. Statistical significance (i.e., number of standard deviations) of the 1, 3, and 5 s binning counting rates observed by the vertical Tupi telescope, as a function of the time elapsed since the MAXI transient 580727270 trigger time.

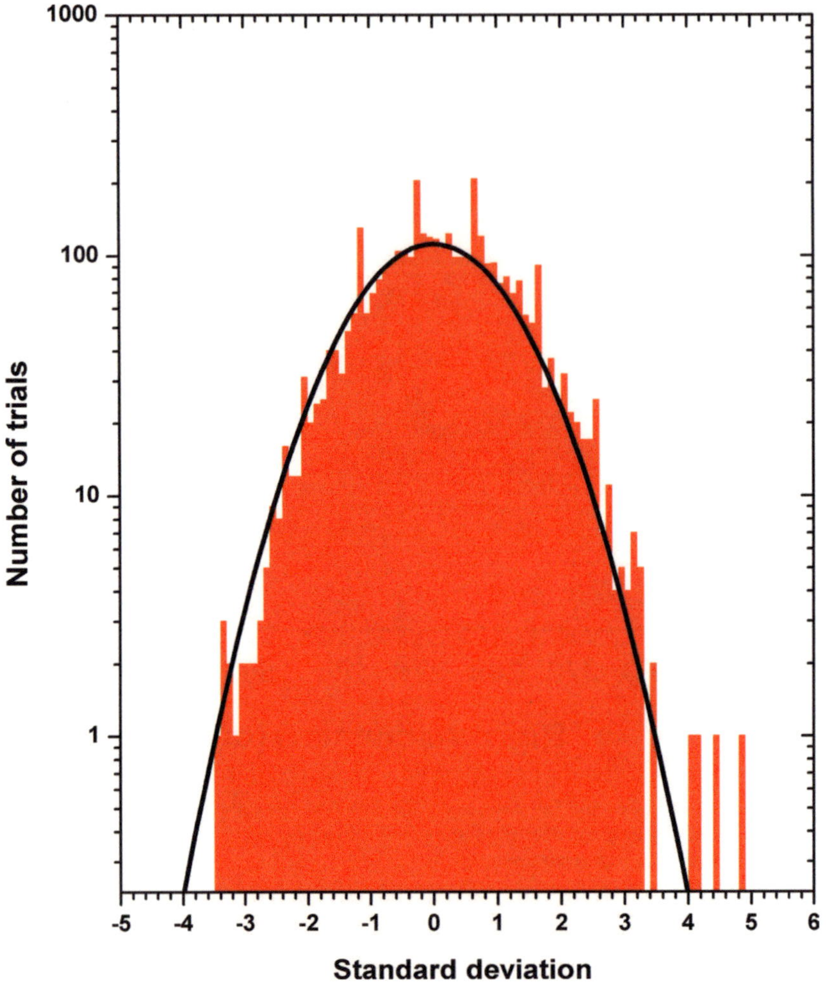

Figure 6. Distribution of the fluctuation count rate for the Tupi telescope (in units of standard deviations) within a temporal windows of 30 min around the MAXI transient event (trigger 580727270). The solid curve represents a Gaussian distribution (background fluctuation) and the signals associated with the MAXI events with significance above 4 sigma clearly are outside of the Gaussian distribution.

We estimated the Poisson probability of the counting rate excess observed in the vertical Tupi telescope, in association with the MAXI GRB events, being a background fluctuation, as $P = (1.6 \pm 0.2) \times 10^{-9}$, i.e., an annual rate of 2.9.

In addition, from spectral analysis, the fluence of the first peak (at To + 25.7 s) was estimated as $F = (2.1 \pm 0.4) \times 10^{-7}$ erg/cm^2.

5.2. Association with the Swift gamma-ray bursts

According to Pagani et al. (GCN 16249), on 12 May 2014 at 19:31:49 UT, the Swift BAT triggered and located a multipeak event with a total duration of about 170 s cataloged as GRB140512A

http://dx.doi.org/10.5772/67176

(trigger = 598819). This Swift event is also in temporal association with the Fermi gamma-ray burst monitor event (Stanbro, GCN 16262). The calculated location by Swift-BAT was (R.A., decl.) = (289.371, −15.100).

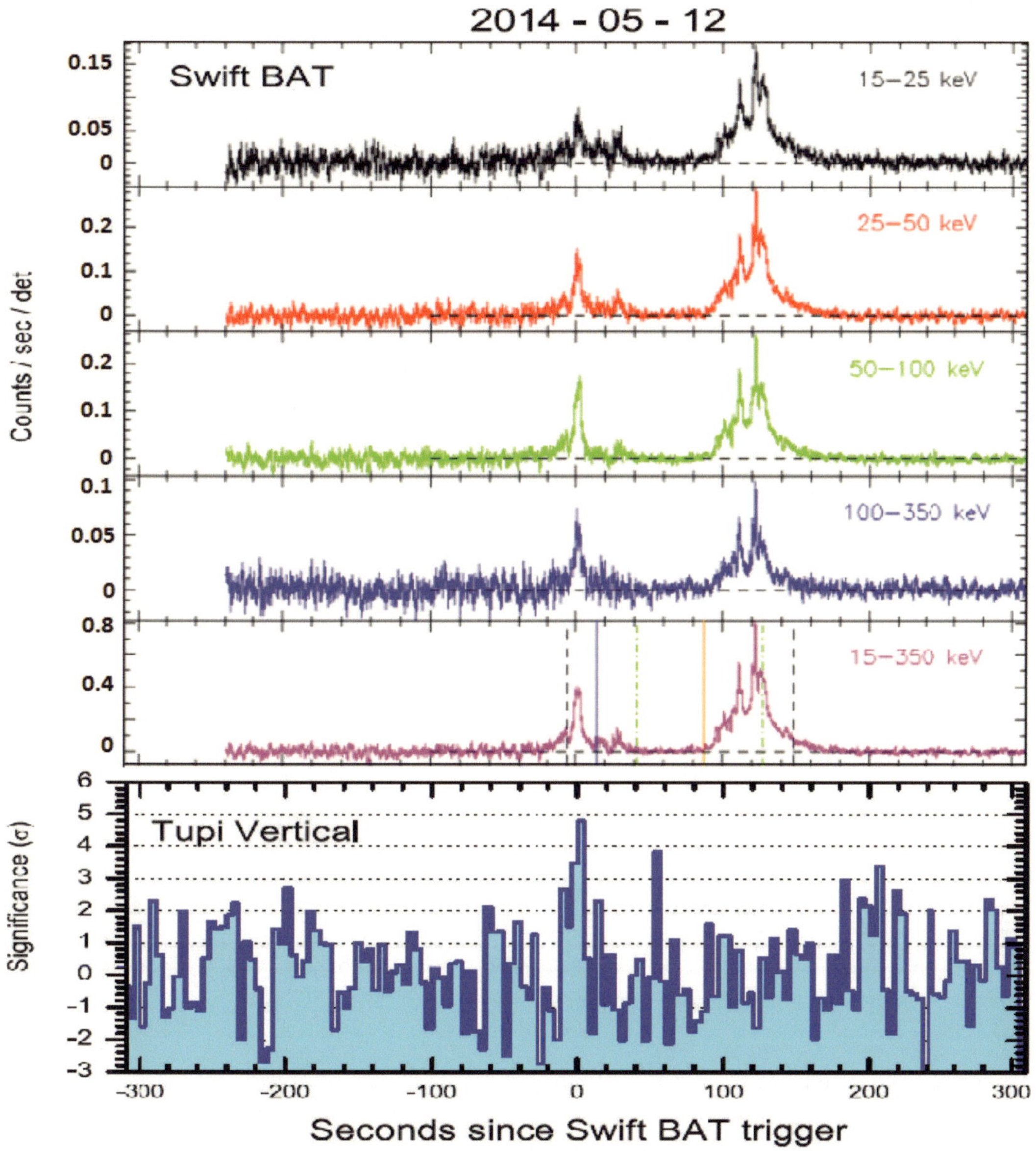

Figure 7. Top panels: the counting rate of gamma rays for five energy ranges for the event Swift BAT GRB140512A. Bottom panel: time profile of the counting rate 4 s binning, and expressed as the number of standard deviations, observed in the Tupi vertical telescope as a function of the time elapsed since the Swift BAT GRB140512A trigger time.

An excess in the Tupi counting rate with a significance of 4.55 sigma was found, temporally and spatially associated with the Swift BAT GRB140512A. The trigger coordinates of this event were very close to the zenith of the Tupi location, that is, within the FOV of the vertical telescope. The signal at Tupi is within the T90 duration of the Swift GRB140512A. **Figure 7** summarizes the situation, where a comparison between the time profiles of Swift BAT and Tupi is shown, as a function of the time elapsed since the Swift BAT GRB140512A trigger time.

The peak in the time profiles of Tupi associated with the Swift GRB140512A persist with the same confidence in the 1, 3, 5, and 10 s binning, as shown in **Figure 8**. This means that the peak is not subjected to be the only background fluctuation.

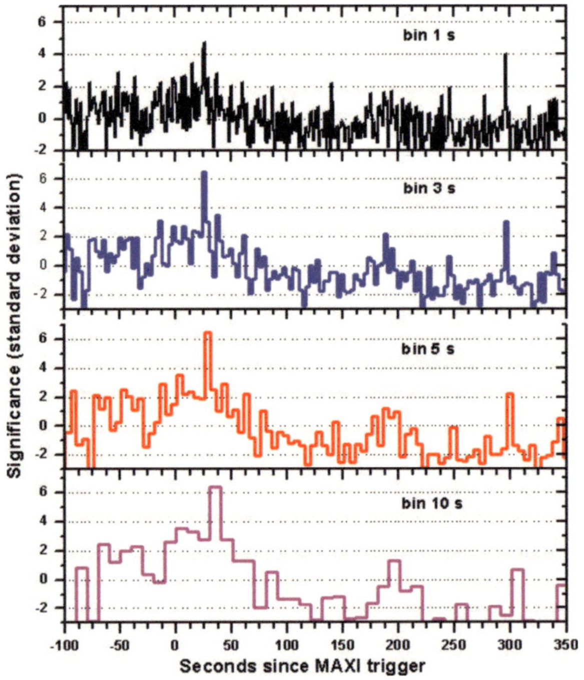

Figure 8. Time profiles observed by the Tupi vertical telescope and expressed as the statistical significance (i.e., number of standard deviations) as a function of the time, since the Swift GRB140512A trigger time and for 1, 3, 5, and 10 s binning. The yellow band marks the region surrounding the Swift trigger time.

To see the expected background fluctuations, a confidence analysis was performed for a 1 h interval around the Swift BAT trigger time, as shown in **Figure 9**. The excess above the Gaussian curve (at right) is linked to the Tupi telescope's signal, associated with the Swift GRB event.

We also estimated the Poisson probability of the excess observed in the counting rate in the vertical Tupi telescope, in association with the Swift GRB event, being a background fluctuation, as $P = (7.40 \pm 1.21) \times 10^{-6}$, i.e., an annual rate of 73.1.

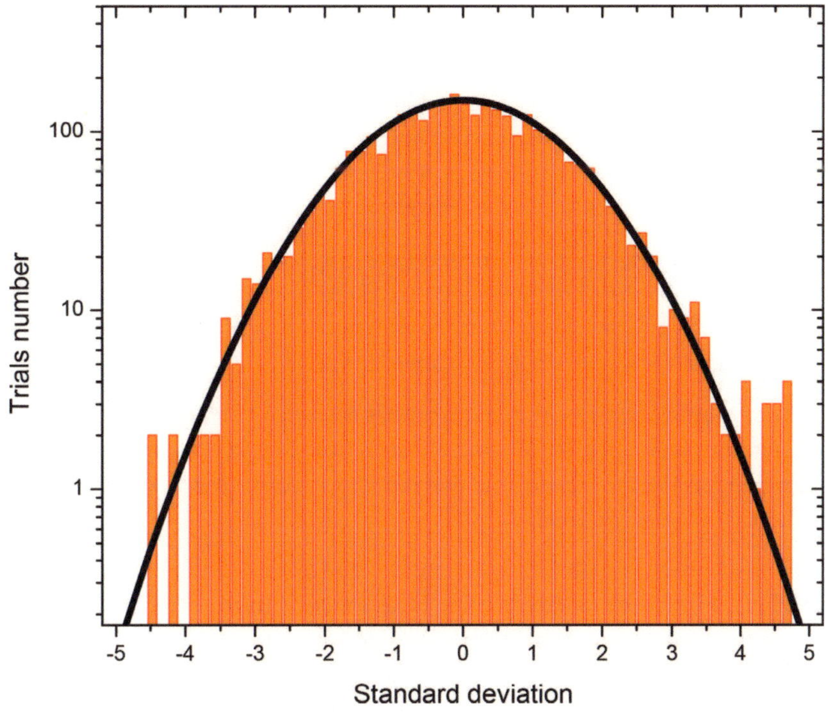

Figure 9. Distribution of the fluctuation count rate for the Tupi telescope (in units of standard deviations) within a temporal windows of 30 min around the Swift GRB140512A transient event. The solid curve represents a Gaussian distribution (background fluctuation) and the signals associated with the Swift events, with significance above 4 sigma clearly are outside of the Gaussian distribution.

6. Summary

We have carried out a systematic search for a GeV counterpart observed at ground level of GRBs triggered in gamma-ray detectors onboard of satellite. An overview on the gamma rays from space was given and the main features observed in detectors onboard satellites, since their discovery in the 1960s and their most recent observations. A brief report is also presented on the main possible mechanisms, as the fireball and the cannonballs models. Both can reproduce the main features of the observed bursts, irrespective of the detailed physics of the central engine.

In addition, several scenarios have been indicated to explain a possible high-energy component of GRBs, such as the synchrotron selfCompton model and the second-order inverse Compton component of the GRB spectrum.

We also included a chronological description of the various efforts for detecting at ground level, the high energy component, that is, the GeV to TeV counterpart of the GRBs. In most cases, ground level detectors have an high energy detection threshold of the secondary particles detected (above 100 GeV); it has not allowed the detection at ground level.

We highlight that the location of the Tupi detector is within the South Atlantic Anomaly (SAA); it allows to achieve higher sensitivity, and some candidates to the GeV counterpart of gamma ray bursts, observed by Tupi telescopes were presented. They are in correlation with temporal and spatial GRBs detected by satellites.

Of course, that the Tupi detector has recorded excess in the counting rate, in correlation with the gamma-ray bursts. As the detector is located at sea level, it is expected that this excess is principally produced by muons. However, the assumption of photomuons as the origin of the excess requires gamma rays with energies above 10 GeV.

In addition, there is an alternative mechanism that is useful to explain high-energy electrons from terrestrial gamma-ray flashes [45] and observations of gamma-ray bursts at ground level under thunderclouds [46]. However, this mechanism requires some special conditions, such as an atmospheric high electric field.

The mechanism is known as "Relativistic runaway electron avalanche in the atmosphere". An initial energetic electron is needed to start the process. In the atmosphere, such energetic electrons typically come from cosmic rays; for instance, gamma rays via pair production process $\gamma \rightarrow e^{+} + e^{-}$ in the upper atmosphere. In this case, there are several seeds for the generation of the successive avalanches; if the atmospheric electric field region is large enough, the number of second-generation avalanches (i.e., avalanches produced by avalanches) will exceed the number of first-generation avalanches, and the number of avalanches itself grows exponentially. This avalanche of avalanches can produce extremely large populations of energetic electrons [47].

Clearly, more studies are needed in order to establish whether this mechanism has the potential to explain the excesses observed in the counting rate of the Tupi detector associated with GRBs. So far, the mechanism "Relativistic runaway electron avalanche in the atmosphere" is the only promising one.

The implications of these ground level observations show that GRBs of long duration have chances of having a GeV counterpart; this characteristic can be useful to the formulations of possible mechanisms of a GeV emission. The experiment is in progress, and the aim is to obtain a large number of candidates in the next years to obtain some systemic features of this phenomenon.

Acknowledgements

The support from National Council for Research (CNPq) and Fundacao de Amparo a Pesquisa do Estado do Rio de Janeiro (FAPERJ) both in Brazil is gratefully acknowledged. We also express our gratitude to V. Kopenkin, C. R. A. Augusto, and A. Nepomuceno for their help in the analysis and to the Goddard Space Flight Center (NASA) for the free access to data through GCN web page (http://gcn.gsfc.nasa.gov/).

Author details

Carlos Navia* and Marcel Nogueira de Oliveira

*Address all correspondence to: tupi.carlos24@gmail.com

Physical Institute, Universidade Federal Fluminense, Niterói, Brazil

References

[1] Lucas J. Live Science. What Are Gamma-Rays? [Internet]. March 20, 2015. Available from: http://www.livescience.com/50215-gamma-rays.html#sthash.KJwQSmKL.dpuf [Accessed: July 15, 2016].

[2] Gardner E., Lattes C. M. G. Production of mesons by the 184-inch Berkeley cyclotron. Science. 1948;**107**(2776):270–271. DOI: 10.1126/science.107.2776.270

[3] Burfening J., Gardner E., Lattes C. M. G. Positive mesons produced by the 184-inch Berkeley cyclotron. Physical Review. 1949;**75**(3):382. DOI: 10.1103/PhysRev.75.382

[4] Kraushaar W. L., Clark G. W. Search for primary cosmic gamma rays with the Satellite Explorer XI. Physical Review Letters. 1962;**8**(3):106. DOI: 10.1103/PhysRevLett.8.106

[5] SPACE.com Staff and NASA. NASA's Top 10 Gamma-Ray Sources in the Universe. [Internet]. December 6, 2011. Available from: http://www.space.com/13838-nasa-gamma-ray-targets-blazars-fermi.html [Accessed: July 15, 2016].

[6] Acciari V. A., Arlen T., Aune T., Benbow W., Bird R., Bouvier A., et al. Observation of Markarian 421 in TeV gamma rays over a 14-year time span. Astroparticle Physics. 2014;**54**:1–10. DOI: 10.1016/j.astropartphys.2013.10.004

[7] Klebesadel R. W., Strong I. B., Olson R. A. Observations of gamma-ray bursts of cosmic origin. Astrophysical Journal. 1973;**182**:L85. DOI: 10.1086/181225

[8] Gehrels N., Chincarini G., Giommi P., Mason K. O., Nousek J. A., Wells A. A., et al. The swift gamma-ray burst mission. The Astrophysical Journal. 2004;**611**(2):1005–1020. DOI: 10.1086/422091

[9] Amati L. Intrinsic spectra and energetics of BeppoSAX gamma-ray bursts with known redshifts. Astronomy & Astrophysics. 2002;**390**(1):81–89. DOI: 10.1051/0004-6361:20020722

[10] Rees M. J., Mészáros P. Relativistic fireballs: energy conversion and time-scales. Monthly Notices of the Royal Astronomical Society. 1992;**258**(1):41P–43P. DOI: 10.1093/mnras/258.1.41P

[11] Rees M. J., Meszaros P. Unsteady outflow models for cosmological gamma-ray bursts. The Astrophysical Journal. 1994;**430**(2):L93–L96. DOI: 10.1086/187446

[12] Cavallo G., Rees M. J. A qualitative study of cosmic fireballs and γ-ray bursts. Monthly Notices of the Royal Astronomical Society. 1978;**183**(3):359–365. DOI: 10.1093/mnras/183.3.359

[13] Paczynski B. Gamma-ray bursters at cosmological distances. The Astrophysical Journal. 1986;**308**:L43–L46. DOI: 10.1086/184740

[14] Shemi A., Piran T. The appearance of cosmic fireballs. The Astrophysical Journal. 1990;**365**:L55–L58. DOI: 10.1086/185887

[15] Wang X. Y., Dai Z. G., Lu T. The inverse Compton emission spectra in the very early afterglows of gamma-ray bursts. The Astrophysical Journal. 2001;**556**(2):1010. DOI: 10.1086/321608

[16] Zhang B., Meszaros P. High-energy spectral components in gamma-ray burst afterglows. The Astrophysical Journal. 2001;**559**(1):110–122. DOI: 10.1086/322400

[17] Pe'er A., Waxman E. The high-energy tail of GRB 941017: comptonization of synchrotron self-absorbed photons. The Astrophysical Journal Letters. 2004;**603**(1):L1–L4. DOI: 10.1086/382872

[18] Sommer M., Bertsch D. L., Dingus B. L., Fichtel C. E., Fishman G. J., Harding A. K., et al. High-energy gamma rays from the intense 1993 January 31 gamma-ray burst. The Astrophysical Journal. 1994;**442**:L63–L66. DOI: 10.1086/187213

[19] Abdo A. A., Ackermann M., Ajello M., Atwood W. B., Axelsson M., Baldini L., et al. Fermi/large area telescope bright gamma-ray source list. The Astrophysical Journal Supplement Series. 2009;**183**(1):46–66. DOI: 10.1088/0067-0049/183/1/46

[20] Spada M., Panaitescu A., Meszaros P. Analysis of temporal features of gamma-ray bursts in the internal shock model. The Astrophysical Journal. 2000;**537**(2):824–832. DOI: 10.1086/309048

[21] Kumar P., McMahon E. A general scheme for modelling γ-ray burst prompt emission. Monthly Notices of the Royal Astronomical Society. 2008;**384**(1):33–63. DOI: 10.1111/j.1365-2966.2007.12621.x

[22] Racusin J. L., Karpov S. V., Sokolowski M., Granot J., Wu X. F., Pal'Shin V., et al. Broadband observations of the naked-eye big gamma-ray burst GRB 080319B. Nature. 2008;**455**(7210):183–188. DOI: 10.1038/nature07270

[23] Mirabel I. F., Rodriguez L. F. Sources of relativistic jets in the Galaxy. Annual Review of Astronomy and Astrophysics. 1999;**37**:409–443. DOI: 10.1146/annurev.astro.37.1.409

[24] Rodriguez L. F., Mirabel I. F. Repeated relativistic ejections in GRS 1915+ 105. The Astrophysical Journal. 1999;**511**(1):398–404. DOI: 10.1086/306642

[25] Dar A. Fireball and cannonball models of gamma-ray bursts confront observations. Chinese Journal of Astronomy and Astrophysics. 2006;**6**(S1):301–314. DOI: 10.1088/1009-9271/6/S1/39

[26] De Rújula A. A unified model of high-energy astrophysical phenomena. International Journal of Modern Physics A. 2005;**20**(29):6562–6583. DOI: 10.1142/S0217751X05029617

[27] Abdo A. A., Ackermann M., Arimoto M., Asano K., Atwood W. B., Axelsson M., et al. Fermi observations of high-energy gamma-ray emission from GRB 080916C. Science. 2009;**323**(5922):1688–1693. DOI: 10.1126/science.1169101

[28] Abdo A. A., Ackermann M., Ajello M., Asano K., Atwood W. B., Axelsson M., et al. Fermi observations of GRB 090902B: a distinct spectral component in the prompt and delayed emission. The Astrophysical Journal Letters. 2009;**706**(1):L138–L144. DOI: 10.1088/0004-637X/706/1/L138

[29] Atkins R., Benbow W., Berley D., Chen M. L., Coyne D. G., Dingus B. L., et al. The high-energy gamma-ray fluence and energy spectrum of GRB 970417a from observations with Milagrito. The Astrophysical Journal. 2003;**583**(2):824–832. DOI: 10.1086/345499

[30] Abdo A. A., Allen B., Berley D., Casanova S., Chen C., Coyne D. G., et al. TeV gamma-ray sources from a survey of the galactic plane with Milagro. The Astrophysical Journal Letters. 2007;**664**(2):L91–L94. DOI: 10.1086/520717

[31] Abdo A. A., Allen B., Berley D., Blaufuss E., Casanova S., Chen C., et al. Discovery of TeV gamma-ray emission from the cygnus region of the galaxy. The Astrophysical Journal Letters. 2007;**658**(1):L33–L36. DOI: 10.1086/513696

[32] Abdo A. A., Allen B. T., Berley D., Blaufuss E., Casanova S., Dingus B. L., et al. Milagro constraints on very high energy emission from short-duration gamma-ray bursts. The Astrophysical Journal. 2007;**666**(1):361–367. DOI: 10.1086/519763

[33] Bartoli B., Bernardini P., Bi X. J., Branchini P., Budano A., Camarri P., et al. Search for GeV gamma-ray bursts with the ARGO-YBJ detector: summary of eight years of observations. The Astrophysical Journal. 2014;**794**(1):82. DOI: 10.1088/0004-637X/794/1/82

[34] Vernetto S. Search for Gamma Ray Bursts at Chacaltaya, arXiv:astro-ph/0011241v1. 2000

[35] Amenomori M., Cao Z., Dai B. Z., Ding L. K., Feng Y. X., Feng Z. Y., et al. Search for 10 TeV burst-like events coincident with the BATSE bursts using the Tibet air shower array. Astronomy and Astrophysics. 1996;**311**:919–926. http://adsabs.harvard.edu/abs/1996A%26A...311..919 T adsnote: Provided by the SAO/NASA Astrophysics Data System

[36] Padilla L., Funk B., Krawczynski H., Contreras J. L., Moralejo A., Aharonian F., et al. Search for gamma-ray bursts above 20 TeV with the HEGRA AIROBICC Cherenkov array. Astronomy and Astrophysics. 1998;**337**:43–50. http://adsabs.harvard.edu/abs/1998A&A...337...43P adsnote: Provided by the SAO/NASA Astrophysics Data System

[37] Poirier J., D'Andrea C., Fragile P. C., Gress J., Mathews G. J., Race D. Search for sub-TeV gamma rays in coincidence with gamma ray bursts. Physical Review D. 2003;**67**(4):042001. DOI: 10.1103/PhysRevD.67.042001

[38] Allard D., Allekotte I., Alvarez C., Asorey H., Barros H., Bertou X., et al. Use of water-Cherenkov detectors to detect gamma ray bursts at the large aperture GRB observatory (LAGO). Nuclear Instruments and Methods in Physics Research Section A: Accelerators, Spectrometers, Detectors and Associated Equipment. 2008;**595**(1):70–72. DOI: 10.1016/j.nima.2008.07.041

[39] Albert J., Aliu E., Anderhub H., Antoranz P., Armada A., Asensio M., et al. Flux upper limit on gamma-ray emission by GRB 050713a from magic telescope observations. The Astrophysical Journal Letters. 2006;**641**(1):L9–L12. DOI: 10.1086/503767

[40] Augusto C. R. A., Navia C. E., Shigueoka H., Tsui K. H., Fauth A. C. Muon excess at sea level from solar flares in association with the Fermi GBM spacecraft detector. Physical Review D. 2011;**84**(4):042002. DOI: 10.1103/PhysRevD.84.042002

[41] NASA. GCN: The Gamma-ray Coordinates Network (TAN: Transient Astronomy Network) [Internet]. Available from: gcn.gsfc.nasa.gov [Accessed: July 19, 2016].

[42] Augusto C. R. A., Navia C. E., De Oliveira M. N., Tsui K. H., Nepomuceno A. A., Kopenkin V., et al. Observation of Muon excess at ground level in relation to gamma-ray bursts detected from space. The Astrophysical Journal. 2015;**805**(1):69. DOI: 10.1088/0004-637X/805/1/69

[43] Ackermann M., Ajello M., Asano K., Axelsson M., Baldini L., Ballet J., et al. The first Fermi-LAT gamma-ray burst catalog. The Astrophysical Journal Supplement Series. 2013;**209**(1):11. DOI: 10.1088/0067-0049/209/1/11

[44] Matsuoka M., Kawasaki K., Ueno S., Tomida H., Kohama M., Suzuki M., et al. The MAXI mission on the ISS: science and instruments for monitoring all-sky X-ray images. Publications of the Astronomical Society of Japan. 2009;**61**(5):999–1010. DOI: 10.1093/pasj/61.5.999

[45] Dwyer J. R., Smith D. M. A comparison between Monte Carlo simulations of runaway breakdown and terrestrial gamma-ray flash observations. Geophysical Research Letters. 2005;**32**(22):L22804. DOI: 10.1029/2005GL023848

[46] Kuroda Y., Oguri S., Kato Y., Nakata R., Inoue Y., Ito C., et al. Observation of gamma ray bursts at ground level under the thunderclouds. Physics Letters B. 2016;**758**:286–291. DOI: 10.1016/j.physletb.2016.05.029

[47] Dwyer J. R. A fundamental limit on electric fields in air. Geophysical Research Letters. 2003;**30**(20):2055. DOI: 10.1029/2003GL017781

Chapter 5

Extragalactic Gamma-Ray Background

Houdun Zeng and Li Zhang

Additional information is available at the end of the chapter

http://dx.doi.org/10.5772/67335

Abstract

The origin of the extragalactic gamma-ray background (EGRB) is an important open issue in the gamma-ray astronomy. There are many theories about the origin of EGRB: (1) some truly diffuse processes, such as dark matter (DM) annihilation or decay, which can produce gamma rays; (2) gamma rays produced by energetic particles accelerated through induced shock waves during structure formation of the universe; (3) a lot of unidentified sources, including normal galaxies, starbursts and active galactic nuclei (AGNs), contain a large number of energetic particles and can emit gamma rays. Among various extragalactic sources, blazars including flat spectral radio quasars (FSRQs) and BL Lac objects are one of the most possible sources for EGRB. As continuous accumulation of the data observed by the Fermi Gamma-Ray Space Telescope, it is possible to directly construct gamma-ray luminosity function (GLF) of the blazars involving evolution information. In this chapter, based on the largest clean sample of AGNs provided by Fermi Large Area Telescope (LAT), we mainly study blazar's GLFs and their contribution to EGRB. In our study, we separately construct GLFs of FSRQs and BL Lacs and then estimate the contributions to EGRB, respectively. Further, we discuss the diffuse gamma ray from other astrophysical sources and the other possible origins of the EGRB.

Keywords: blazars, gamma-ray radiation, luminosity function, the extragalactic gamma-rays background

1. Introduction

The large area telescope (LAT [1]) onboard Fermi gamma-ray space telescope (Fermi) has measured the extragalactic diffuse gamma-ray background and then provided useful information for us to study the origins of the extragalactic gamma-ray background (EGRB) [2–5]. However, the origin of the EGRB is still an unsolved problem. Observationally, an isotropic component of the EGRB emission was first detected by the SAS-2 satellite [6, 7] and subsequently measured by the energetic gamma-ray experiment telescope (EGRET) [8–10]. Due to the higher sensitivity of Fermi-LAT than that of EGRET, the observed integrated flux above 100MeV by the LAT is

$(1.03–0.17) \times 10^{-5}$ photons $cm^{-2}s^{-1}$ [3], which is lower than $(1.14–0.05) \times 10^{-5}$ photons $cm^{-2}s^{-1}$ measured by EGRET [11]. Recently, Fermi-LAT has made a new measurement of the EGRB spectrum and their results shown that the EGRB energy spectrum between 0.1 and 820GeV is to be well represented by a power law with an exponential cutoff above 300GeV [5]. **Figure 1** (left panel) shows the measured X-ray and gamma-ray background radiation spectra. We know that the X-ray background spectrum has no big change with time and has been considered as the integrated light produced via the accretion process of active galactic nuclei (AGNs) [12]. However, the gamma-ray spectrum is different from the X-ray background spectrum due to the sensitivity of an instrument and other reasons. Before the Fermi gamma-ray space telescope era, neither spectrum nor origin of the EGRB was well understood. In particular, the spectrum at 0.03–50GeV reported by EGRET has a break in the several GeV. With the arrival of Fermi era, more accurate determination of the EGRB spectrum and more extragalactic source samples are provided to understand the nature of the EGRB. Note that the whole gamma-ray sky contains diffuse galactic emission, point sources, isotropic extragalactic diffuse emission and local and solar diffuse emissions. **Figure 1** (right panel) shows that the EGRB spectrum is obtained by removing the resolved point source, like as the most recent list of resolved Fermi-LAT source (3FGL), the diffuse galactic emission determined by GALPROP, which simulates both cosmic-ray propagation in the galaxy and the gamma-ray flux resulting from interactions and possibly an isotropic flux of galactic, by restricting data to regions with $|b| > 10°$ or even higher galactic latitudes.

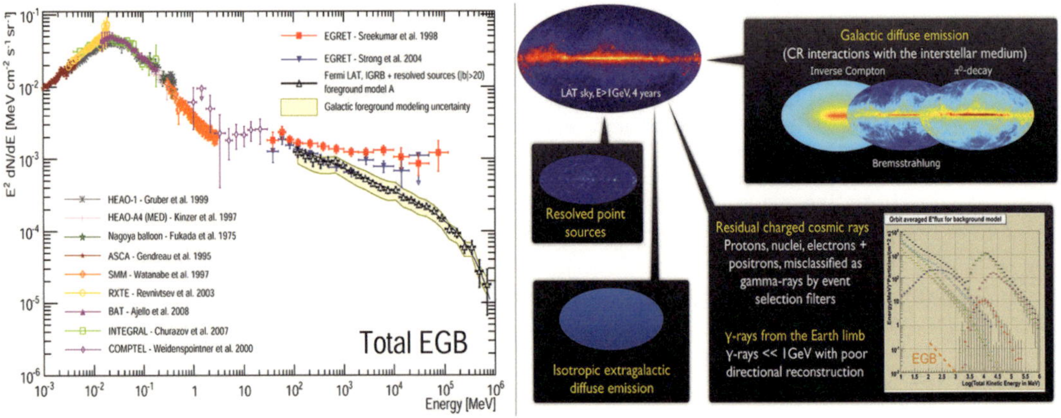

Figure 1. Left: The measured X-ray and gamma-ray background radiation spectra, which is from Ref. [5]. Right: The composition of the total gamma-ray flux. The figure is obtained from the report of Ackermann, M. at Fermi Symposium.

Similar to the extragalactic EGRET sky, blazars are the largest source class identified by Fermi extragalactic sky and their contribution to the EGRB has been widely discussed. Typical estimated contributions of unresolved blazars to the EGRB range from 10 to 100% [13–36]. Blazars are divided into two main subgroups: BL Lac objects and FSRQs [37]. Among the gamma-ray blazar sample, the number of FSRQs detected by Fermi-LAT is smaller than that of BL Lac objects (e.g., 2FGL, 3FGL). FSRQs generally show softer spectrum in the gamma-ray band (e.g., [38]), which is to be detected harder than BL Lac objects at a given significant limit. On the one hand, BL Lacs are reputed as the population of extragalactic sources that show a negative or no cosmological evolution [39–42], but FSRQs are regarded as those with a positive cosmological evolution, which is similar to the population of X-ray-selected, radio-quiet AGNs [43–45]. Ajello et al. [32] suggested that BL Lacs have a more complex evolution. At the modest redshift region, most BL Lac classes show a positive evolution with a space density peaking. Meanwhile, their results suggest that the evolution of low-luminosity, high-synchrotron-peaked (HSP) BL Lac objects is strong negative with number density increasing for low redshift range ($z \leq 0.5$). In addition, the contributions of the EGRB from other sources or processes are very important. Those are star-forming galaxies [46, 47], radio galaxies (e.g., [14, 46, 48]), gamma-ray bursts (GRBs) (e.g., [49]), high galactic-latitude pulsars (e.g., [50]), intergalactic shocks (e.g., [51, 52]), Seyferts (e.g., [53]), cascade from ultra-high-energy cosmic rays (e.g., [54, 55]), large galactic electron halo [56], cosmic-ray interaction in the solar system [57] and dark matter annihilation or decay (e.g., [58]). Recently, with the assumptions and uncertainties, Ajello et al. [33] and Di Mauro and Donato [36] shown that the EGRB can be fully accounted for the sum of contributions from undetected sources including blazars and radio and star-forming galaxies. Those results imply that little room in space is left for other processes such as shock wave or DM interactions (e.g., [33, 59]).

The extragalactic gamma-ray sky provides an amount of gamma-ray sources and allows us to obtain the information about the evolution of sources and estimate their contributions to the EGRB. Because the blazar's contribution is the main content of research on this chapter, the detail about how to build the gamma-ray luminosity function (GLF) will be discussed in Section2. In Section3, a brief description about how to estimate different components' contributions to the EGRB is given and finally, we give the conclusions and discussions in Section4.

2. The gamma-ray luminosity function

Since the Fermi-LAT has detected and identified more and more gamma-ray sources and observed previously detected objects in greater detail, the method by using the gamma-ray luminosity function (GLF) to estimate the EGRB of resolved sources has become much more reliable. In this approach, the GLF involving the evolution of redshift as well as the distribution of spectral indices of a given source class can be established for all known sources and the observed population can be extrapolated to lower fluxes.

2.1. Function derivation

As professed in Ref. [31], there is a classical approach to obtain the luminosity function, which is on account of 1/VMAX method provided by Schmidt [60] to deal with redshift bins. However, this method has a fault, which is known to introduce bias in each binning. For a small sample and/or a large span of parameters, if the bins contained significant evolution, the method would result in a loss of important information. In order to constrain the model parameters for various models of the evolving GLF, a maximum likelihood method is adopted, which is first introduced by Marshall et al. [61]. The likelihood function L is given as follows (e.g., [17, 19, 24, 62]):

$$\mathcal{L} = \exp\left(-N_{\rm exp}\right)\prod_{i=1}^{N_{obs}}\Phi(L_{\gamma,i},\ z_i,\ \Gamma_i), \tag{1}$$

where $N_{\rm exp}$ is the expected number of source detections:

$N_{\rm exp} = \int d\Gamma \int dz \int dL_\gamma \Phi(L_{\gamma,i}, z_i, \Gamma_i)$, $N_{\rm exp}$ is the number of the sample of sources and $\Phi(L_{\gamma,i}, z_i, \Gamma_i)$ is the distribution function of the space density of source on luminosity (L_γ), redshift (z) and photon index (Γ). The function form can be expressed as follows:

$$\Phi(L_{\gamma,i},\ z_i,\ \Gamma_i) = \frac{d^3N}{dL_\gamma dz d\Gamma} = \rho_\gamma(L_\gamma,\ z) \times \frac{dN}{d\Gamma} \times \frac{dV}{dz} \times \omega(L_\gamma,\ z,\ \Gamma), \tag{2}$$

where $\rho_\gamma(L_\gamma, z)$ is the γ-ray luminosity function and dV/dz is the comoving volume element per unit redshift and unit solid angle:

$dV/dz = cd_L^2/(H_0(1+z)^2\sqrt{\Omega_M(1+z)^3 + \Omega_\Lambda})$. $dN/d\Gamma$ is the intrinsic photon index distribution assumed as a Gaussian $\exp(-(\Gamma-\mu)^2/2\sigma^2)$, where μ and σ are the mean and the dispersion, respectively. $\omega(L_\gamma, z, \Gamma)$ is the detection efficiency and represents the probability of detecting an object with the γ-ray luminosity L_γ at redshift z and photon index Γ [1, 24, 31]. The relationship between χ^2 and likelihood (L) can be expressed by function $\chi^2 = -2 \ln (L)$ [63]. In this case, the function $\chi^2 = -2 \ln L$ that is minimized is defined as follows:

$$\chi^2 = -2\sum_i^{N_{obs}} \ln(\Phi(L_{\gamma,i}, z_i, \Gamma_i)) + 2N_{\rm exp}. \tag{3}$$

For a given GLF, the redshift distribution, luminosity distribution and photon index distribution can be divided into three intervals of size $dL_\gamma dz d\Gamma$ and the three kinds of differential distributions can be expressed from GLF as follows [31]:

$$\begin{aligned} \frac{dN}{dz} &= \int_{\Gamma_{min}}^{\Gamma_{max}} \int_{L_{\gamma,min}}^{L_{\gamma,max}} \frac{d^3N}{dL_\gamma dz d\Gamma} dL_\gamma dz, \\ \frac{dN}{dz} &= \int_{\Gamma_{min}}^{\Gamma_{max}} \int_{z_{min}}^{z_{max}} \frac{d^3N}{dL_\gamma dz d\Gamma} dz d\Gamma, \\ \frac{dN}{dz} &= \int_{\Gamma_{min}}^{\Gamma_{max}} \int_{L_{\gamma,min}}^{L_{\gamma,max}} \frac{d^3N}{dL_\gamma dz d\Gamma} dL_\gamma dz, \end{aligned} \tag{4}$$

The source count distribution can be derived as follows:

$$\begin{aligned} \mathrm{N}\,(>S) &= \int_{\Gamma_{\rm min}}^{\Gamma_{\rm max}} d\Gamma \int_{z_{\rm min}}^{z_{\rm max}} dz \int_{L_\gamma(z,S)}^{L_{\gamma,\ \rm max}} dL_\gamma \frac{d^3N}{dL_\gamma dz d\Gamma} \\ &= \int_{\Gamma_{\rm min}}^{\Gamma_{\rm max}} \frac{dN}{d\Gamma} d\Gamma \int_{z_{\rm min}}^{z_{\rm max}} \frac{dV}{dz} \int_{L_\gamma(z,S)}^{L_{\gamma,\ \rm max}} \rho_\gamma(L_\gamma, z)\omega(L_\gamma,\ z,\ \Gamma) dL_\gamma \end{aligned} \tag{5}$$

where $L_\gamma(z, S)$ is the luminosity of a source at redshift z with a flux of S_γ (>100MeV).

Through minimized Eq. (3), we can obtain the best-fitting parameters of the models. There are multiple parameters in our various models to find the best in observational data in a multidimensional model parameter space; the MCMC technique can be employed for its high efficiency to constrain the model parameters. In this method, the Metropolis-Hastings algorithm that generates samples from the posterior distribution using a Markov Chain is used when sampling the model parameters and the probability density distributions of the model parameters are asymptotically proportional to the number density of the sample points. For each parameter set P, one obtains the likelihood function $L(P) \propto \exp\left(-\chi^2(P)/2\right)$, where χ^2 is obtained by comparing model predictions with observations. A new set of parameter $P^{'}$ is adopted to replace the existing one P with a probability of min $\{1, L(P^{'})/L(P)\}$. The MCMC method has been reviewed by Fan et al.[64] and described in detail by Neal [65], Gamerman [66], Lewis and Bridle [67], Mackay [68].

2.2. Models description

The GLF models for different source classes are uncertainty. Currently, there are two methods for constructing the blazars' GLF: the first method is to build the GLF by assuming a relationship between the GLF and the luminosity function in a lower energy band, for example, that the GLF relates to radio luminosity function (RLF) or to the X-ray luminosity function (XLF) (e.g., [14, 16, 17, 19, 23, 28, 48, 69–72]); the second method is to construct the GLF directly using observed gamma-ray data of blazars (e.g., [15, 17, 22]). Before the Fermi era, constructing the GLF model indirectly was used more frequently due to the small EGRET samples, which results in blazar's contribution between the range of 10 and 100%. In next sections, we briefly review those models for directly constructing the GLF.

2.2.1. The pure density evolution

The pure density evolution (PDE) model is the simplest scenario of evolution and the GLF has a following form:

$$\rho(L_\gamma, z) = \frac{A}{\ln(10)L_\gamma}\left[\left(\frac{L_\gamma}{L}\right)^{\gamma_1} + \left(\frac{L_\gamma}{L}\right)^{\gamma_2}\right]^{-1} \times e(z), \tag{6}$$

where $e(z) = (1+z)^\kappa$ is the standard power-law evolutionary factor. In this case, there are five model parameters and other two parameters, μ and σ, are also added.

2.2.2. The pure luminosity evolution

In the pure luminosity evolution (PLE) model, the GLF can be expressed as follows:

$$\rho(L_\gamma,\ z) = \frac{A(1+z)^\kappa e^{z/\xi}}{\ln(10)L_\gamma}\left[\left(\frac{L_\gamma}{L_*(1+z)^\kappa e^{z/\xi}}\right)^{\gamma_1} + \left(\frac{L_\gamma}{L_*(1+z)^\kappa e^{z/\xi}}\right)^{\gamma_2}\right]^{-1}, \tag{7}$$

where A is a normalization factor, L_i is the evolving break luminosity, γ_1 is the faint-end slope index, γ_2 is the bright-end slope index, κ and ξ represent the redshift evolution. Including the parameters μ and σ, there are 8 parameters in calculations.

2.2.3. The luminosity-dependent density evolution

In the luminosity-dependent density evolution (LDDE) model, the GLF evolution is decided by a redshift cutoff that depends on luminosity and the GLF can be given by

$$\rho(L_\gamma, z) = \frac{A}{\ln(10)L_\gamma}\left[\left(\frac{L_\gamma}{L_*}\right)^{\gamma_1} + \left(\frac{L_\gamma}{L_*}\right)^{\gamma_2}\right]^{-1}$$
$$\left[\left(\frac{1+z}{1+z_c^*(L_\gamma/10^{48})^\alpha}\right)^{p_1} + \left(\frac{1+z}{1+z_c^*(L_\gamma/10^{48})^\alpha}\right)^{p_2}\right], \tag{8}$$

where A is a normalization factor, L is evolving break luminosity, γ_1 and p_1 are the faint-end slope index, γ_2 and p_2 are the bright-end slope index, z_c is redshift peak with a luminosity (here 10^{48} ergs s^{-1}) and α is power-law index of the redshift-peak evolution. From this, there are 10 parameters for calculation.

The detailed description about PLE and LDDE models can be found in sections4.1 and 4.2 from Ref. [32]. These models also can be applied to X-ray band, to determine the information of evolution of sources in X-ray band (e.g., [62]). With the increase in the number of the detected sources, the evolutionary form of those sources becomes more complicated and the updated forms of those models can be found in Ref. [33], which allows the Gaussian mean μ of the photon index and the evolutionary factor $e(z, L_\gamma)$ to change with luminosity.

2.3. The cosmological evolution

In Fermi sample, the large redshift range between $z = 0$ and $z = 3.1$ of gamma-ray blazars was found. The obtained GLFs have shown that blazars have a cosmological evolution in their gamma-ray band. We have simply discussed the redshift evolution of blazars in the "Introduction". Ajello et al. [32] recently have presented the new results on the cosmological evolution of the BL Lac population by using the largest and most complete sample of gamma-ray BL Lacs available in the literature and they found that for most BL Lac classes, the evolution is positive, with a space density peaking at modest redshift (z≈1.2) (see **Figure 2**). In **Figure 2**, we also see that for their higher luminosity, FSRQs dominate at all redshifts z>0.3 and the extreme growth in BL Lac numbers at low z allows them to produce ~90% of the local luminosity density. In particular, low-luminosity, high-synchrotron-peaked (HSP) BL Lac objects showed different evolutionary behaviors with respect to other blazar classes (see **Figure 2**). They have strong

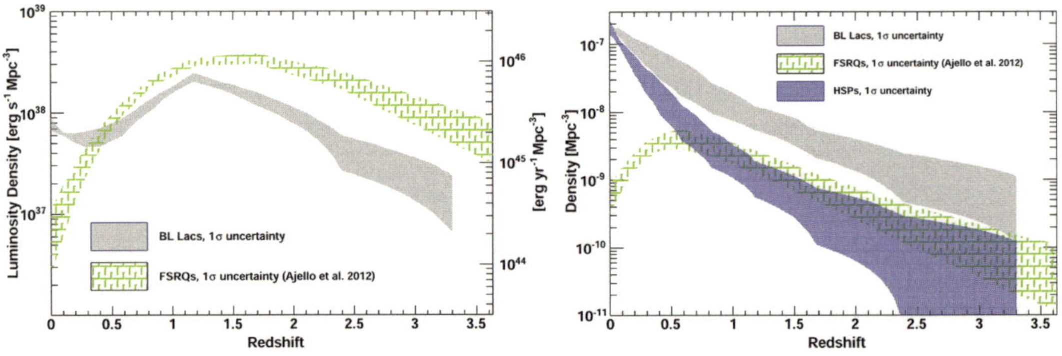

Figure 2. Left: The evolution of the luminosity density of FSRQs compared to that of BL Lac objects. Right: Number density of FSRQs, BL Lac objects and HSPs. The figures are obtained from the report of [32] and see Ref. [32] for additional details.

negative evolution with number density increasing for z<0.5, which confirms previous standpoints of negative evolution based on the samples of X-ray-selected BL Lac objects and this sample contained a large fraction of HSPs [39, 41].

3. The extragalactic gamma-ray background

The origin of the EGRB has been widely discussed for various gamma-ray-emitting sources in literature. Fermi has observed gamma-ray emission from blazars, star-forming galaxies, radio galaxies, GRBs and high-latitude pulsars. Ajello et al. [33] and Di Mauro and Donato [36] suggested that blazars, star-forming galaxies and radio galaxies are the main contributors to the EGRB. For those emitting sources, we focus on how to estimate the contribution of unresolved objects to the EGRB below, based on the best-fitting GLF (space density of sources).

The differential intensity of the EGRB radiation can be expressed as follows:

$$\frac{dN}{dEd\Omega}=\int_{\Gamma_{min}}^{\Gamma_{max}} d\Gamma\frac{dN}{d\Gamma}\int_{z_{min}}^{z_{max}}\frac{dV}{dz}\int_{L_{\gamma,min}}^{L_{\gamma,max}} dL_{\gamma}$$
$$\Phi(L_{\gamma},z)F_{\gamma}^{intrinsic}(E,L_{\gamma},z,\Gamma)e^{-\tau(E,z)}\left(1.0-\omega(L_{\gamma},z,\Gamma)\right) \qquad (9)$$

where $\Phi(L_{\gamma},z)$ is the GLF and $e^{-\tau(E,z)}$ is the optical depth of the extragalactic background light (EBL) for the sources at redshift z emitting gamma-ray photon energy E. Recently, there are many studies on EBL (e.g., [21, 73–75]). Generally, we adopted the model given by [73] for the EBL to calculate the optical depth. In Eq. (9), $F_{\gamma}^{intrinsic}(E,L_{\gamma},z,\Gamma)$ represents the intrinsic photon flux at energy E with γ-ray luminosity L_{γ} and a power-law spectrum at redshift z and it is expressed as follows:

$$F_{\gamma}^{\mathrm{intrinsic}}(E,\ L_{\gamma},\ z,\ \Gamma)=\frac{L_{\gamma}\ (1+z)^{2-\Gamma}}{4\pi d_L^2 E_1^2}\begin{cases}(2-\Gamma)\left[\left(\frac{E_2}{E_1}\right)^{2-\Gamma}-1\right]^{-1}\left(\frac{E}{100\ \mathrm{MeV}}\right)^{-\Gamma} & \Gamma\neq 2,\\ \frac{1}{\ln(E_2/E_1)}\left(1-\frac{E_1}{E_2}\right)^{-1}\left(\frac{E}{100\ \mathrm{MeV}}\right)^{-2} & \Gamma=2,\end{cases} \qquad (10)$$

where $E_1=100$ MeV and $E_2=100$ GeV. Therefore, the integrated intensity between photon energy E_1 and E_2 ($E_2>E_1$) can be written as follows:

$$\frac{dN}{d\Omega}=\int_{E_1}^{E_2}\frac{dN}{dEd\Omega}dE \qquad (11)$$

The electrons and positrons are produced due to the interaction between very high energy (VHE) photons from TeV sources and ultraviolet-infrared photons of EBL. The pairs could scatter the cosmic microwave background (CMB) radiation to high-energy background radiation through the inverse Compton scattering process (e.g., [76–83]). This cascading emission is regarded as a contributor to the EGRB if the flux of the cascade flux is lower than the detector's sensitivity. Now, we consider only the first generation of the electron-positron pairs produced by the gamma-ray absorption to obtain the cascade emission because the emission from the second generation or more than second generation of created pairs can be negligible at the GeV band [21]. The formulation of the cascade flux is given as follows [84]:

$$F_{\gamma}^{\text{cascade}}(E,\ L_{\gamma},\ z,\ \Gamma) = \frac{81\ \pi}{16\ \lambda_c^3} \frac{\varepsilon_c^2 m_e c^2}{(1+z)^4 U_{\text{CMB}}} \int_{\sqrt{3\varepsilon_c/4\varepsilon_{\text{CMB}}(1+z)}}^{\infty} \frac{d\gamma}{\gamma^8 \exp[3\varepsilon_c/4\gamma^2 \varepsilon_{\text{CMB}}(1+z) - 1]} \times \int_{2\gamma}^{\varepsilon_{\max}} d\varepsilon\ F_{\text{VHE}}^{\text{intrinsic}}\left(\frac{5.11\times10^5}{10^6}\varepsilon,\ z,\ L_{\gamma},\ \Gamma\right)[1 - e^{-\tau(\varepsilon,z)}] \tag{12}$$

where $\lambda_c = 2.426\times10^{-10}$ cm is the Compton length, the dimensionless energy $\varepsilon_c = E\times10^6/(5.11\times10^5)$, $U_{CMB} = 4.0\times10^{-13}$ erg cm^{-3} is the CMB energy density at $z = 0.0$, $\varepsilon_{CMB} = 1.24\times10^9 m_e c^2$ is the average CMB photon energy at $z = 0.0$ and $\varepsilon_{CMB} = 2.0\times10^8$ corresponding to $E_{VHE} = 100 TeV$. $F_{VHE}^{intrinsic}(E_{VHE}, L_\gamma, z, \Gamma)$ represents the possible intrinsic TeV spectrum, which is extrapolated to the TeV energy ranges from the observed GeV spectrum Eq. (10) by assuming a power-law spectrum. In Eq. (9), using Eq. (12) in place of Eq. (10) allows us to compute the contribution to the EGRB from the cascade emission of the source.

It is noted that there are two possible contributions for the cascade emission to the EGRB because the pairs are deflected by the extragalactic magnetic field (EGMF), which is shown in **Figure 3**. In case I, the cascade emission can contribute to the EGRB if the flux of the cascade emission is lower than that of the LAT sensitivity. In case II, although the flux of the cascade emission is larger than that of the LAT sensitivity, the angle between the redirected secondary gamma-ray photons and the line of sight is larger than that of the LAT point-spread function (PSF) (i.e., $\theta > \theta_{PSF}$). Thus, the cascade emission will not be attributed to a point source by the LAT and it then contributes to the EGRB, where $\theta_{PSF} = (1.7\pi/180)(0.001E)^{-0.74}[1 + (0.001E/15)^2]^{0.37}$ [85]. For more detailed information, see Refs. [81, 84].

3.1. Blazars

Blazars emit gamma rays via the inverse Compton scattering processes and/or hadronic processes and dominate extragalactic gamma-ray sources. Therefore, it is naturally expected that blazars contribute the main EGRB. However, its fraction was very uncertain in the EGRET era due to its small samples. At the same time, its fraction also severely depends on GLF. Blazars are divided into two main subgroups: BL Lac objects and FSRQs [37]. **Figure 4** shows FSRQs' EGRB spectra with LDDE model and BL Lacs' EGRB spectra with PDE model. Compared to FSRQs, BL Lacs have lower gamma-ray luminosities, lower redshifts and harder spectral indices in statistics (e.g., [86]). Thus, BL Lacs can provide a significant part in the contribution of blazar to the EGRB above 10GeV. From **Figure 4**, we find out that the cascade emission from BL Lacs has a rather large fraction of the total EGRB energy flux and contrary to that of FSRQs, which may be caused by harder spectrum for BL Laces. Therefore, the contribution from BL Lacs cascade emission to the EGRB cannot be negligible. Based on the effect of the EGMF on the cascade contribution from blazars, Yan [84] have studied the effect of cascade radiation on the contribution to the EGRB using a simple semi-analytical model. They suggested that if the strength of the EGMF is large enough ($B_{EGMF} > 10^{-12}$ G), the cascade contribution can significantly alter the spectrum of the EGRB at high energies. If the small strength of the EGMF is large enough ($B_{EGMF} < 10^{-14}$ G), then the cascade contribution is small, but it cannot be ignored. Recently, Ajello et al. [33] used an updated GLFs to analyze the redshift, luminosity and photon index distributions and obtained the best-fitting evolutionary parameters of the GLFs. According to the GLFs and spectral energy distribution (SED) model consistent with the Fermi blazar observations, their result shown that blazars account for 50^{+12}_{-10} to the EGRB (see **Figure 5**).

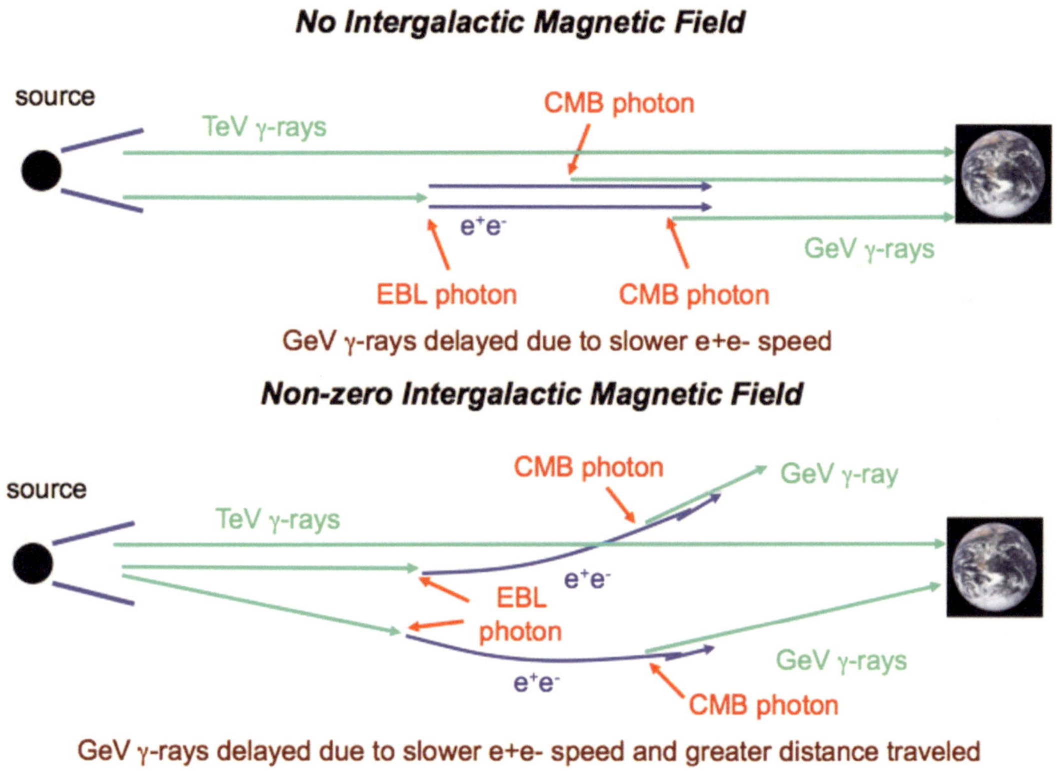

Figure 3. The cascade radiation processes in no or non-zero extragalactic magnetic field (EGMF). Note that the pairs produced by the interaction between very high energy (VHE) photons and ultraviolet-infrared photons of EBL are detected by the EGMF. The figures are obtained from the report of Marco Ajello at Fermi Symposium.

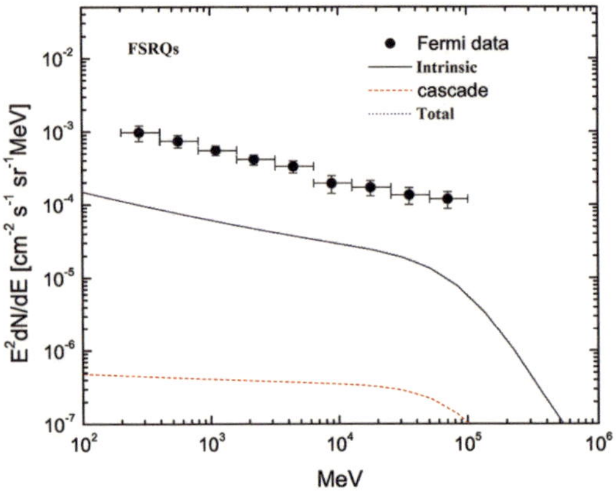

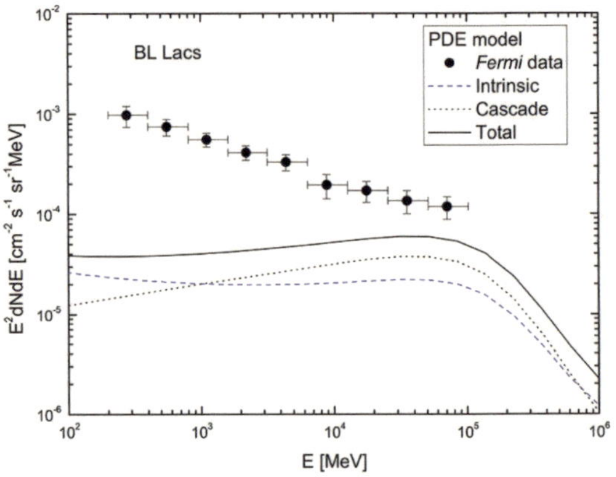

Figure 4. Comparison of predicted EGRB spectra from FSRQs and BL Lacs with the observed data of blazars. Note that the EGRB spectra from FSRQs and BL Lac are estimated based on LDDE and PDE models, respectively. The two figures are obtained from the report of Refs. [29, 30].

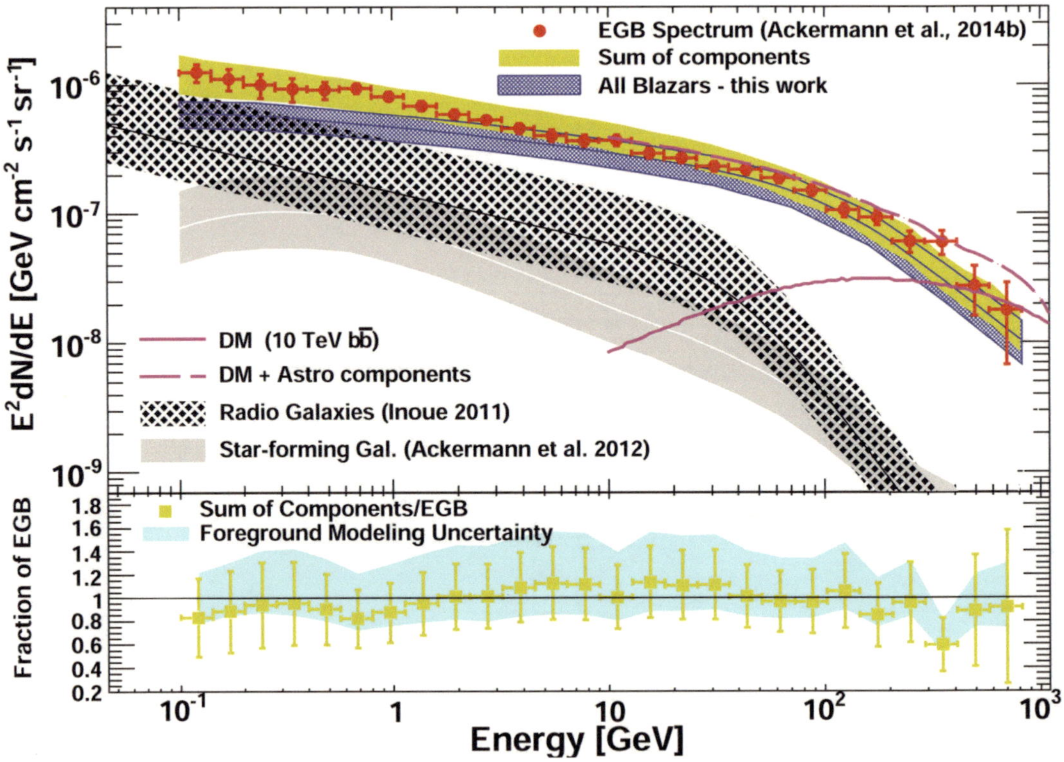

Figure 5. The EGRB spectrum of blazars [33], star-forming galaxies (gray band [4]) and radio galaxies (black striped band [48]) as well as summation of these three populations (yellow band), compared to the intensity of the observed ERGB [5]. The figure is obtained from the report of Ref. [33].

3.2. Radio galaxies

Radio galaxies are one of the largest subclasses of radio-loud AGNs. It is more in number than blazars in the entire sky. Even though radio galaxies are fainter than blazars, Fermi-LAT has just detected gamma rays from ~15 extragalactic sources, including 12 FR Is and 3 FR IIs [87]. In order to estimate the contribution to the EGRB from radio galaxies, their GLF is required. We must obtain indirectly the GLF due to the limited Fermi radio galaxy samples. Relying on a correlation between the luminosities in the radio and gamma-ray frequencies, Inoue [48] converted the RLF [88] into the GLF and estimated about 25% of EGRB can be solved by radio galaxies (see **Figure 5**). This uncertainty significantly depends on the limited sample and the errors between the gamma ray and radio luminosity correlation.

3.3. Star-forming galaxies

The Fermi-LAT has detected gamma-ray from ~9 star-forming (SF) galaxies [2]. Those gamma rays are produced by interactions between cosmic rays and gas or interstellar radiation fields, including the decay of neutral pion and electron interactions (bremsstrahlung and inverse Compton scattering). Similar to radio galaxies, it is not straightforward to construct the GLF because of the limited star-forming galaxy sample. Generally, the correlations between the IR wavelength and gamma-ray region are used to predict the gamma-ray diffuse emission for the unresolved SF galaxy population. Different from other types of source, the SF gamma-ray average spectrum is difficult to firmly establish due to the paucity of statistics. Milky Way-like SF galaxies (MW model) and an assumed power-law spectrum (PL model) are proposed by Ackermann et al. [89] to express an average spectrum of SF Galaxies. In particular, the two predictions are different above 5GeV, where the MW model softens significantly. Therefore, using the correlation between infrared and gamma-ray luminosities, based on the well-established infrared luminosity functions and the SF gamma-ray average spectrum, the GLF of star-forming galaxies is well built and the contribution of star-forming galaxies to the EGRB can be estimated as 10–30% of the EGRB at >0.1GeV [89], which can be seen in **Figure 5**.

It should be noted that about 95% of the EGRB can be naturally explained by blazars, star-forming galaxies and radio galaxies in the 0.1–820GeV range. Only modest space is left for other diffuse processes such as dark matter interactions, which suggests that other gamma-ray-emitting sources' contribution can be neglected. Ajello et al. [33] also concluded that the result of their simulation gave an upper limit on DM self-annihilation cross sections, which is similar to that from the independent types of analysis (e.g., [59]).

4. Conclusion and discussion

In this chapter, we reviewed the origin of EGRB and estimated the contribution of unsolved gamma-ray-emitting sources from Fermi-LAT to the EGRB based on the construction of the corresponding GLFs. Since Fermi-LAT has higher sensitivity and provides numerous gamma-ray-emitting sources for studies, we found two important results: (i) the redshift evolutionary information of gamma-ray sources, particularly for blazars; HBLs show strong negative cosmological evolution, while FSRQs and luminous BL Lacs show positive evolution like as Seyferts and the cosmic star formation history. (ii) Fermi sources' contribution to the EGRB; blazars clearly contribute to most of the EGRB (≈40-62%), as well as radio galaxies and star-forming galaxies can occupy for the rest room of the EGRB [33, 36]. These results suggest that the contributions of other emitting sources have only little space to the EGRB. However, the uncertainties associated with these predictions from radio galaxies and star-forming galaxies are still quite large because of the small samples. This situation is very similar to blazar studies in the early EGRET era. Therefore, further data will be required to construct the GLFs and precisely evaluate the contributions from those two populations.

Now, there are still some unresolved problems. We have not seen the signature of dark matter particles in the EGRB spectrum, although they are considered as the possible origin of EGRB. As we known, Fermi-LAT has accurately measured the EGRB spectrum and the anisotropy of the EGRB [4] and the emission from dark matter is anisotropic and its spatial pattern is unique and predictable [90]. Therefore, we can obtain an upper limit on the annihilation cross section by comparing the expected EGRB angular power spectrum from dark matter annihilation with the measured spectrum. The work of Ajello et al. [33] shown that an analysis of the EGRB and its components can constrain diffuse emission mechanisms such as DM annihilation. Di Mauro and Donato [36] probed a possible emission coming from the annihilation of WIMP DM in the halo of our galaxy and found that the DM component can very well fit the EGRB data together with the realistic emission from a number of unresolved extragalactic sources.

The value of the EGMF has still not been determined. Since the pairs scatter CMB photons to GeV energies by Compton mechanism for cascade process around a TeV sources, Fermi-LAT could measure those GeV photons, which would give a straight measurement of the EGMF. As continuous accumulation of the data observed and the further development of detection equipment, the imprint of the EGMF may be found in the gamma-ray spectrum and/or flux [79, 80, 91]. The EGMF imprint might also be found in the angular anisotropy of the EGRB [92]. If the effect of cascade depending on the EGMF cannot be neglected [84], the electron-positron pairs produced in cascade process could be deflected by a high value of the EGMF, which makes GeV photons more isotropic. Therefore, the EGRB spectrum with the anisotropy could probe the strength of EGMF [87].

Author details

Houdun Zeng[1,3] and Li Zhang[2,3]*

*Address all correspondence to: lizhang@ynu.edu.cn

1 Key Laboratory of Dark Matter and Space Astronomy, Purple Mountain Observatory, Chinese Academy of Sciences, Nanjing, China

2 Department of Astronomy, Yunnan University, Kunming, China

3 Key Laboratory of Astroparticle Physics of Yunnan Province, Kunming, China

References

[1] Atwood, W. B.; Abdo, A. A.; Ackermann, M.; Althouse, W.; Anderson, B.; et al. The large area telescope on the Fermi gamma-ray space telescope mission. The Astrophysical Journal. 2009;**697**(2):1071–1102. doi:10.1088/0004-637X/697/2/1071

[2] Abdo, A. A.; Ackermann, M.; Ajello, M.; Allafort, A.; Antolini, E.; et al. The first catalog of active galactic nuclei detected by the Fermi large area telescope. The Astrophysical Journal. 2010;**715**(1): 429–457. doi:10.1088/0004-637X/715/1/429

[3] Abdo, A. A.; Ackermann, M.; Ajello, M.; Atwood, W. B.; Baldini, L.; Ballet, J.; et al. Spectrum of the isotropic diffuse gamma-ray emission derived from first-year Fermi large area telescope data. Physical Review Letters. 2010;**104**(10):101101. doi:10.1103/PhysRevLett.104.101101

[4] Ackermann, M.; Ajello, M.; Albert, A.; Baldini, L.; Ballet, J.; Barbiellini, G.; et al. Anisotropies in the diffuse gamma-ray background measured by the Fermi LAT. Physical Review D. 2012;**85**(8):083007. doi:10.1103/PhysRevD.85.083007

[5] Ackermann, M.; Ajello, M.; Albert, A.; Atwood, W. B.; Baldini, L.; Ballet, J.; et al. The spectrum of isotropic diffuse gamma-ray emission between 100 MeV and 820 GeV. The Astrophysical Journal. 2015;**799**(1):86, 24 pp. doi:10.1088/0004-637X/799/1/86

[6] Fichtel, C. E.; Simpson, G. A.; Thompson, D. J. Diffuse gamma radiation. The Astrophysical Journal. 1978;**222**(1):833–849. doi:10.1086/156202

[7] Thompson, D. J.; Fichtel, C. E. Extragalactic gamma radiation—use of galaxy counts as a galactic tracer. Astronomy and Astrophysics. 1982;**109**(2):352–354.

[8] Osborne, J. L.; Wolfendale, A. W.; Zhang, L. The diffuse flux of energetic extragalactic gamma rays. Journal of Physics G: Nuclear and Particle Physics. 1994;**20**(7):1089–1101. doi:10.1088/0954-3899/20/7/010

[9] Willis, T. D., Observations of the Isotropic Diffuse Gamma-Ray Background with the EGRET Telescope, Ph.D. thesis, Stanford University, Aug 1996

[10] Sreekumar, P.; Bertsch, D. L.; Dingus, B. L.; Esposito, J. A.; Fichtel, C. E.; et al. EGRET observations of the extragalactic gamma-ray emission. The Astrophysical Journal. 1998;**494**(2):523–534. doi:10.1086/305222

[11] Strong, A. W.; Moskalenko, I. V.; Reimer, O. A new determination of the extragalactic diffuse gamma-ray background from EGRET data. The Astrophysical Journal. 2004;**613**(2):956–961. doi:10.1086/423196

[12] Ueda, Y.; Akiyama, M.; Hasinger, G.; Miyaji, T.; Watson, M. G. Toward the standard population synthesis model of the x-ray background: evolution of x-ray luminosity and absorption functions of active galactic nuclei including compton-thick populations. The Astrophysical Journal. 2014;**786**(2):104, 28 pp. doi:10.1088/0004-637X/786/2/104

[13] Stecker, F. W.; Salamon, M. H.; Malkan, M. A. The high-energy diffuse cosmic gamma-ray background radiation from blazars. The Astrophysical Journal. 1993;**410**(2):L71–L74. doi:10.1086/186882

[14] Padovani, P.; Ghisellini, G.; Fabian, A. C.; Celotti, A. Radio-loud AGN and the extragalactic gamma-ray background. Monthly Notices of the Royal Astronomical Society. 1993;**260**(3): L21–L24. doi:10.1093/mnras/260.1.L21

[15] Chiang, J.; Fichtel, C. E.; von Montigny, C.; Nolan, P. L.; Petrosian, V. The evolution of gamma-ray--loud active galactic nuclei. The Astrophysical Journal. 1995;**452**:165. doi:10.1086/176287

[16] Stecker, F. W.; Salamon, M. H. The gamma-ray background from blazars: a new look. The Astrophysical Journal. 1996;**464**:600. doi:10.1086/177348

[17] Chiang, J.; Mukherjee, R. The luminosity function of the EGRET gamma-ray blazars. The Astrophysical Journal. 1998;**496**(2):752–760. doi:10.1086/305403

[18] Mücke, A.; Pohl, M. The contribution of unresolved radio-loud AGN to the extragalactic diffuse gamma-ray background. Monthly Notices of the Royal Astronomical Society. 2000;**312**(1):177–193. doi:10.1046/j.1365-8711.2000.03099.x

[19] Narumoto, T.; Totani, T. Gamma-ray luminosity function of blazars and the cosmic gamma-ray background: evidence for the luminosity-dependent density evolution. The Astrophysical Journal. 2006;**634**(1):81–91. doi:10.1086/502708

[20] Dermer, C. D. Statistics of cosmological black hole jet sources: blazar predictions for the gamma-ray large area space telescope. The Astrophysical Journal. 2007;**659**(2):958–975. doi:10.1086/512533

[21] Kneiske, T. M.; Mannheim, K. BL Lacertae contribution to the extragalactic gamma-ray background. Astronomy and Astrophysics. 2008;**479**(1):41–47. doi:10.1051/0004-6361:20065605

[22] Bhattacharya, D.; Sreekumar, P.; Mukherjee, R. Gamma-ray luminosity function of gamma-ray bright AGNs. Research in Astronomy and Astrophysics. 2008;**9**(1):85–94. doi:10.1088/1674-4527/9/1/007

[23] Inoue, Y.; Totani, T. The blazar sequence and the cosmic gamma-ray background radiation in the Fermi era. The Astrophysical Journal. 2009;**702**(1):523–536. doi:10.1088/0004-637X/702/1/523

[24] Abdo, A. A.; Ackermann, M.; Ajello, M.; Antolini, E.; Baldini, L.; Ballet, J.; et al. The Fermi-LAT high-latitude survey: source count distributions and the origin of the extragalactic diffuse background. The Astrophysical Journal. 2010;**720**(1):435–453. doi:10.1088/0004-637X/720/1/435

[25] Ghirlanda, G.; Ghisellini, G.; Tavecchio, F.; Foschini, L.; Bonnoli, G. The radio-γ-ray connection in Fermi blazars. Monthly Notices of the Royal Astronomical Society. 2011;**413**(2):852–862. doi:10.1111/j.1365-2966.2010.18173.x

[26] Stecker, F. W.; Venters, T. M. Components of the extragalactic gamma-ray background. The Astrophysical Journal. 2011;**736**(1):40, 13 pp. doi:10.1088/0004-637X/736/1/40

[27] Singal, J.; Petrosian, V.; Ajello, M. Flux and photon spectral index distributions of Fermi-LAT blazars and contribution to the extragalactic gamma-ray background. The Astrophysical Journal. 2012;**753**(1):45, 11 pp. doi:10.1088/0004-637X/753/1/45

[28] Zeng, H. D.; Yan, D. H.; Sun, Y. Q.; Zhang, L. γ-Ray luminosity function and the contribution to extragalactic γ-ray background for Fermi-detected blazars. The Astrophysical Journal. 2012;**749**(2):151, 8 pp. doi:10.1088/0004-637X/749/2/151

[29] Zeng, H.; Yan, D.; Zhang, L. A revisit of gamma-ray luminosity function and contribution to the extragalactic diffuse gamma-ray background for Fermi FSRQs. Monthly Notices of the Royal Astronomical Society. 2013;**431**(1):997–1003. doi:10.1093/mnras/stt223

[30] Zeng, H.; Yan, D.; Zhang, L. Gamma-ray luminosity function of BL Lac objects. Monthly Notices of the Royal Astronomical Society. 2014;**441**(2):1760–1768. doi:10.1093/mnras/stu644

[31] Ajello, M.; Shaw, M. S.; Romani, R. W.; Dermer, C. D.; Costamante, L.; et al. The luminosity function of Fermi-detected flat-spectrum radio quasars. The Astrophysical Journal. 2012;**751**(2):108, 20 pp. doi:10.1088/0004-637X/751/2/108

[32] Ajello, M.; Romani, R. W.; Gasparrini, D.; Shaw, M. S.; Bolmer, J.; et al. The cosmic evolution of Fermi BL lacertae objects. The Astrophysical Journal. 2014;**780**(1):73, 24 pp. doi:10.1088/0004-637X/780/1/73

[33] Ajello, M.; Gasparrini, D.; Sánchez-Conde, M.; Zaharijas, G.; Gustafsson, M.; et al. The origin of the extragalactic gamma-ray background and implications for dark matter annihilation. The Astrophysical Journal Letters. 2015;**800**(2): L27, 7 pp. doi:10.1088/2041-8205/800/2/L27

[34] Di Mauro, M.; Donato, F.; Lamanna, G.; Sanchez, D. A.; Serpico, P. D. Diffuse γ-ray emission from unresolved BL lac objects. The Astrophysical Journal. 2014;**786**(2):129, 12 pp. doi:10.1088/0004-637X/786/2/129

[35] Di Mauro, M.; Calore, F.; Donato, F.; Ajello, M.; Latronico, L.. Diffuse γ-ray emission from misaligned active galactic nuclei. The Astrophysical Journal. 2014;**780**(2):161, 14 pp. doi:10.1088/0004-637X/780/2/161

[36] Di Mauro, M.; Donato, F. Composition of the Fermi-LAT isotropic gamma-ray background intensity: emission from extragalactic point sources and dark matter annihilations. Physical Review D. 2015;**91**(12):123001. doi:10.1103/PhysRevD.91.123001

[37] Urry, C. Megan; Padovani, Paolo. Unified schemes for radio-loud active galactic nuclei. Publications of the Astronomical Society of the Pacific. 1995;**107**:803. doi:10.1086/133630

[38] Acero, F.; Ackermann, M.; Ajello, M.; Albert, A.; Atwood, W. B.; Axelsson, M.; et al. Fermi large area telescope third source catalog. The Astrophysical Journal Supplement Series. 2015;**218**(2):23, 41 pp. doi:10.1088/0067-0049/218/2/23

[39] Rector, Travis A.; Stocke, John T.; Perlman, Eric S.; Morris, Simon L.; Gioia, Isabella M. The properties of the x-ray-selected EMSS sample of BL lacertae objects. The Astronomical Journal. 2000;**120**(4):1626–1647. doi:10.1086/301587

[40] Caccianiga, A.; Maccacaro, T.; Wolter, A.; Della Ceca, R.; Gioia, I. M. On the cosmological evolution of BL lacertae objects. The Astrophysical Journal. 2002;**566**(1):181–186. doi:10.1086/338073

[41] Beckmann, V.; Engels, D.; Bade, N.; Wucknitz, O. The HRX-BL Lac sample—evolution of BL lac objects. Astronomy and Astrophysics. 2003;**401**:927–938. doi:10.1051/0004-6361:20030184

[42] Padovani, P.; Giommi, P.; Landt, H.; Perlman, Eric S. The deep X-ray radio blazar survey. III. Radio number counts, evolutionary properties and luminosity function of blazars. The Astrophysical Journal. 2007;**662**(1):182–198. doi:10.1086/516815

[43] Dunlop, J. S.; Peacock, J. A. The redshift cut-off in the luminosity function of radio galaxies and quasars. Monthly Notices of the Royal Astronomical Society. 1990;**247**(1):19.

[44] Ueda, Y.; Akiyama, M.; Ohta, K.; Miyaji, T. Cosmological evolution of the hard x-ray active galactic nucleus luminosity function and the origin of the hard x-ray background. The Astrophysical Journal. 2003;**598**(2):886–908. doi:10.1086/378940

[45] Hasinger, G.; Miyaji, T.; Schmidt, M. Luminosity-dependent evolution of soft X-ray selected AGN. New Chandra and XMM-Newton surveys. Astronomy and Astrophysics. 2005;**441**(2):417–434. doi:10.1051/0004-6361:20042134

[46] Strong, A. W.; Wolfendale, A. W.; Worrall, D. M.. Origin of the diffuse gamma ray background. Monthly Notices of the Royal Astronomical Society. 1976;**175**:23–27. doi:10.1093/mnras/175.1.23P

[47] Pavlidou, V.; Fields, B. D. The guaranteed gamma-ray background. The Astrophysical Journal. 2002;**575**(1):L5–L8. doi:10.1086/342670

[48] Inoue, Y. Contribution of gamma-ray-loud radio galaxies' core emissions to the cosmic MeV and GeV gamma-ray background radiation. The Astrophysical Journal. 2011;**733**(1):66, 9 pp. doi:10.1088/0004-637X/733/1/66

[49] Casanova, S.; Dingus, B. L.; Zhang, B. Contribution of GRB emission to the GeV extragalactic diffuse gamma-ray flux. The Astrophysical Journal. 2007;**656**(1):306–312. doi:10.1086/510613

[50] Faucher-Giguère, Claude-André; Loeb, Abraham. The pulsar contribution to the gamma-ray background. Journal of Cosmology and Astroparticle Physics. 2010;**01**(01):005. doi:10.1088/1475-7516/2010/01/005

[51] Loeb, A.; Waxman, E. Cosmic γ-ray background from structure formation in the intergalactic medium. Nature. 2000;**405**(6783):156–158.

[52] Totani, T.; Kitayama, T. Forming clusters of galaxies as the origin of unidentified GEV gamma-ray sources. The Astrophysical Journal. 2000;**545**(2):572–577. doi:10.1086/317872

[53] Inoue, Y.; Totani, T.; Ueda, Y. The cosmic MeV gamma-ray background and hard x-ray spectra of active galactic nuclei: implications for the origin of hot AGN coronae. The Astrophysical Journal Letters. 2008;**672**(1): L5. doi:10.1086/525848

[54] Dar, A.; Shaviv, N. J. Origin of the high energy extragalactic diffuse gamma ray background. Physical Review Letters. 1995;**75**(17):3052–3055. doi:10.1103/PhysRevLett.75.3052

[55] Kalashev, O. E.; Semikoz, D. V.; Sigl, G. Ultrahigh energy cosmic rays and the GeV-TeV diffuse gamma-ray flux. Physical Review D. 2009;**79**(6): 063005. doi:10.1103/PhysRevD.79.063005

[56] Keshet, U.; Waxman, E.; Loeb, A. The case for a low extragalactic gamma-ray background. Journal of Cosmology and Astroparticle Physics. 2004;**04**:006. doi:10.1088/1475-7516/2004/04/006

[57] Moskalenko, I. V.; Porter, T. A. Isotropic gamma-ray background: cosmic-ray-induced albedo from debris in the solar system? The Astrophysical Journal Letters. 2009;**692**(1): L54–L57. doi:10.1088/0004-637X/692/1/L54

[58] Bergström, L.; Edsjö, J.; Ullio, P. Spectral gamma-ray signatures of cosmological dark matter annihilations. Physical Review Letters. 2001;**87**(25):251301. doi:10.1103/PhysRevLett.87.251301

[59] Ackermann, M.; Albert, A.; Anderson, B.; Baldini, L.; Ballet, J.; Barbiellini, G.; Bastieri, D.; Bechtol, K.; Bellazzini, R.; Bissaldi, E.; and 112 coauthors. Dark matter constraints from observations of 25 MilkyÂ Way satellite galaxies with the Fermi large area telescope. Physical Review D. 2014;**89**(4):042001. doi:10.1103/PhysRevD.89.042001

[60] Schmidt, M. Space distribution and luminosity functions of quasi-stellar radio sources. The Astrophysical Journal. 1968;**151**:393. doi:10.1086/149446

[61] Marshall, H. L.; Tananbaum, H.; Avni, Y.; Zamorani, G.. Analysis of complete quasar samples to obtain parameters of luminosity and evolution functions. The Astrophysical Journal. 1983;**269**(1):35–41. doi:10.1086/161016

[62] Ajello, M.; Costamante, L.; Sambruna, R. M.; Gehrels, N.; Chiang, J.; Rau, A.; et al. The evolution of Swift/BAT Blazars and the origin of the MeV background. The Astrophysical Journal. 2009;**699**(1):603–625. doi:10.1088/0004-637X/699/1/603

[63] Kochanek, C. S. The flat-spectrum radio luminosity function, gravitational lensing, galaxy ellipticities and cosmology. The Astrophysical Journal. 1996;**473**:595. doi:10.1086/178175

[64] Fan, Z. H.; Liu, S. M.; Yuan, Q.; Fletcher, L. Lepton models for TeV emission from SNR RX J1713.7–3946. Astronomy and Astrophysics. 2010;**517**:L4, 4 pp. doi:10.1051/0004-6361/201015169

[65] Neal, R. M. Probabilistic Inference Using Markov Chain Monte Carlo Methods, Technical Report. Department of Computer Science, Univ.Toronto: 1993

[66] Gamerman, D. Markov Chain Monte Carlo: Stochastic Simulation for Bayesian Inference. Chapman and Hall: London; 1997

[67] Lewis, A.; Bridle, S. Cosmological parameters from CMB and other data: a Monte Carlo approach. Physical Review D. 2002;**66**(10):103511. doi:10.1103/PhysRevD.66.103511

[68] MacKay, D. J. C. Information theory, inference and learning algorithms[M]. Cambridge university press, 2003;P 640, ISBN-13: 978-0521642989

[69] Zhang, L.; Cheng, K. S.; Fan, J. H. The radio and gamma-ray luminosities of blazars. Publications of the Astronomical Society of Japan. 2001;**53**(2):207–213. doi:10.1093/pasj/53.2.207

[70] Narumoto, T.; Totani, T. Gamma-ray luminosity function of blazars and the cosmic gamma-ray background: evidence for the luminosity-dependent density evolution. Astrophysics and Space Science.2007;**309**(1–4):73–79. doi:10.1007/s10509-007-9453-4

[71] Abazajian, K. N.; Blanchet, S.; Harding, J. P. Contribution of blazars to the extragalactic diffuse gamma-ray background and their future spatial resolution. Physical Review D. 2011;**84**(10):103007. doi:10.1103/PhysRevD.84.103007

[72] Li, F.; Cao, X.-W. BL Lacertae objects and the extragalactic γ-ray background. Research in Astronomy and Astrophysics. 2011;**11**(8):879–887. doi:10.1088/1674-4527/11/8/001

[73] Finke, J. D.; Razzaque, S.; Dermer, C. D. Modeling the extragalactic background light from stars and dust. The Astrophysical Journal. 2010;**712**(1):238–249. doi:10.1088/0004-637X/712/1/238

[74] Domínguez, A.; Primack, J. R.; Rosario, D. J.; Prada, F.; Gilmore, R. C.; Faber, S. M.; et al. Extragalactic background light inferred from AEGIS galaxy-SED-type fractions. Monthly Notices of the Royal Astronomical Society. 2011;**410**(4):2556–2578. doi:10.1111/j.1365-2966.2010.17631.x

[75] Inoue, Y.; Inoue, S.; Kobayashi, M. A. R.; Makiya, R.; Niino, Y.; Totani, T. Extragalactic Background Light from hierarchical galaxy formation: gamma-ray attenuation up to the epoch of cosmic reionization and the first stars. The Astrophysical Journal. 2013;**768** (2):197, 17 pp. doi:10.1088/0004-637X/768/2/197

[76] Fan, Y. Z.; Dai, Z. G.; Wei, D. M. Strong GeV emission accompanying TeV blazar H1426 +428. Astronomy and Astrophysics. 2004;**415**:483–486. doi:10.1051/0004-6361:20034472

[77] Murase, K.; Takahashi, K.; Inoue, S.; Ichiki, K.; Nagataki, S. Probing intergalactic magnetic fields in the GLAST era through pair echo emission from TeV blazars. The Astrophysical Journal Letters. 2008;**686**(2): L67. doi:10.1086/592997

[78] Yang, C. Y.; Fang, J.; Lin, G. F.; Zhang, L.. Possible GeV Emission from TeV blazars. The Astrophysical Journal. 2008;**682**(2):767–774. doi:10.1086/589326

[79] Neronov, A.; Vovk, I. Evidence for strong extragalactic magnetic fields from Fermi observations of TeV blazars. Science. 2010;**328**(5974):73. doi:10.1126/science.1184192

[80] Tavecchio, F.; Ghisellini, G.; Foschini, L.; Bonnoli, G.; Ghirlanda, G.; Coppi, P. The intergalactic magnetic field constrained by Fermi/large area telescope observations of the TeV

blazar 1ES0229+200. Monthly Notices of the Royal Astronomical Society: Letters. 2010;**406**(1):L70–L74. doi:10.1111/j.1745-3933.2010.00884.x

[81] Dermer, C. D.; Cavadini, M.; Razzaque, S.; Finke, J. D.; Chiang, J.; Lott, B. Time delay of cascade radiation for TeV blazars and the measurement of the intergalactic magnetic field. The Astrophysical Journal Letters. 2011;**733**(2):L21. doi:10.1088/2041-8205/733/2/L21

[82] Huan, H.; Weisgarber, T.; Arlen, T.; Wakely, S. P.. A new model for gamma-ray cascades in extragalactic magnetic fields. The Astrophysical Journal Letters. 2011;**735**(2):L28, 5 pp. doi:10.1088/2041-8205/735/2/L28

[83] Inoue, Y.; Ioka, K. Upper limit on the cosmological gamma-ray background. Physical Review D. 2012;**86**(2):023003. doi:10.1103/PhysRevD.86.023003

[84] Yan, D.; Zeng, H.; Zhang, L. Contribution from blazar cascade emission to the extragalactic gamma-ray background: what role does the extragalactic magnetic field play? Monthly Notices of the Royal Astronomical Society. 2012;**422**(2):1779–1784. doi:10.1111/j.1365-2966.2012.20752.x

[85] Taylor, A. M.; Vovk, I.; Neronov, A. Extragalactic magnetic fields constraints from simultaneous GeV-TeV observations of blazars. Astronomy & Astrophysics. 2011;**529**:A144, 9 pp. doi:10.1051/0004-6361/201116441

[86] Ackermann, M.; Ajello, M.; Allafort, A.; Antolini, E.; Atwood, W. B.; et al. The second catalog of active galactic nuclei detected by the Fermi large area telescope. The Astrophysical Journal. 2011;**743**(2):171, 37 pp. doi:10.1088/0004-637X/743/2/171

[87] Massaro. F.; Thompson, D. J.; Ferrara, E. C. The extragalactic gamma-ray sky in the Fermi era. The Astronomy and Astrophysics Review. 2016;**24**:1.

[88] Willott, C. J.; Rawlings, S.; Blundell, K. M.; Lacy, M.; Eales, S. A. The radio luminosity function from the low-frequency 3CRR, 6CE and 7CRS complete samples. Monthly Notices of the Royal Astronomical Society. 2001;**322**(3):536–552. doi:10.1046/j.1365-8711.2001.04101.x

[89] Ackermann, M.; Ajello, M.; Allafort, A.; Baldini, L.; Ballet, J.; Bastieri, D.; et al. GeV observations of star-forming galaxies with the Fermi large area telescope. The Astrophysical Journal. 2011;**755**(2):164, 23 pp. doi:10.1088/0004-637X/755/2/164

[90] Ando, S.; Komatsu, E.; Narumoto, T.; Totani, T. Dark matter annihilation or unresolved astrophysical sources? Anisotropy probe of the origin of the cosmic gamma-ray background. Physical Review D. 2007;**75**(6):063519. doi:10.1103/PhysRevD.75.063519

[91] Tavecchio, F.; Ghisellini, G.; Bonnoli, G.; Foschini, L. Extreme TeV blazars and the intergalactic magnetic field. Monthly Notices of the Royal Astronomical Society. 2011;**414**(4):3566–3576. doi:10.1111/j.1365-2966.2011.18657.x

[92] Venters, T. M.; Pavlidou, V. Probing the intergalactic magnetic field with the anisotropy of the extragalactic gamma-ray background. Monthly Notices of the Royal Astronomical Society. 2013;**432**(4):3485–3494. doi:10.1093/mnras/stt697

Section 3

Recent Applications of Gamma Rays

Chapter 6

Applications of Ionizing Radiation in Mutation Breeding

Özge Çelik and Çimen Atak

Additional information is available at the end of the chapter

http://dx.doi.org/10.5772/66925

Abstract

As a predicted result of increasing population worldwide, improvements in the breeding strategies in agriculture are valued as mandatory. The natural resources are limited, and due to the natural disasters like sudden and severe abiotic stress factors, excessive floods, etc., the production capacities are changed per year. In contrast, the yield potential should be significantly increased to cope with this problem. Despite rich genetic diversity, manipulation of the cultivars through alternative techniques such as mutation breeding becomes important. Radiation is proven as an effective method as a unique method to increase the genetic variability of the species. Gamma radiation is the most preferred physical mutagen by plant breeders. Several mutant varieties have been successfully introduced into commercial production by this method. Combinational use of *in vitro* tissue culture and mutation breeding methods makes a significant contribution to improve new crops. Large populations and the target mutations can be easily screened and identified by new methods. Marker assisted selection and advanced techniques such as microarray, next generation sequencing methods to detect a specific mutant in a large population will help to the plant breeders to use ionizing radiation efficiently in breeding programs.

Keywords: mutation breeding, in vitro mutagenesis, gamma rays, molecular markers, high-throughput technologies

1. Introduction

The worldwide population is expected to be nine billion at 2050. Conventional agricultural crops are inadequate to meet the current need to provide sustainable yield production. Therefore, crop improvement is getting an important need when we are not able to meet the demands of growing world population. For this reason, humans have begun to develop new plant varieties for cultivation, and it is called as plant breeding. Numerous food, feed, and ornamental and industrial crops were improved via hybridization methods to meet the needs of human beings since many years. Over the last 15 years, development of new techniques became useful in breeding strategies to facilitate the improvement of new crop varieties.

Plant breeding methods and recent progress in biotechnology contribute greatly to friendly agriculture. The main point is to establish productive breeding strategies to improve crops.

Variation is the main point of the breeding that the plant breeders are focused. Genetic variation is a natural phenomenon. This variation is a natural result of genotypes, which have interactions with the environmental facts, get together. The recombination and independent assortment of the alleles are responsible to obtain new individuals from the population. Domestication of the crops is affected by several conditions such as ecological and agricultural. Selection of the adaptive genotypes is getting important in breeding of the cultivars. The main point is to achieve the production of higher-yielding crops [1], useful traits such as size of the fruits, and quality of the crops. The aim of the breeding is to combine various features of many plants in one plant. This method is general for breeding of the plants via sexual reproduction. During recombination of the alleles, offsprings carrying selectable variations for the several traits exist. Recombination is not responsible to produce new traits itself. Although genetic changes have provided the natural variation for species evolution, changes in species have not only been important for adaptation to natural environment. Mutations are the main reasons of genetic variabilities and cause new species eventually. Therefore, they have also been exploited by man in the agricultural processes of species domestication and crop improvement. As a new approach, manipulation of the cultivars through alternative techniques such as mutation breeding and Biotechnology are useful for especially some fully sterile plants [2].

Mutations have been shown as a way of procreating variations in a variety. They spontaneously occur in nature. Several mistakes can cause mutations during replication process. On the other hand, radiation is an efficient mutagen that the plants are exposed. The important point is the origin of the mutated cell. Somatic cell mutations are not easily traceable and cannot pass to the future generations; otherwise, embryonic cell mutations directly pass to the next springs. Spontaneous mutations occur without any human intervention and happen randomly with a low frequency. However, some mutagenic agents are known to induce mutations as an alternative to low incidences of spontaneous mutations to increase genetic variability by increasing the frequency of mutations. Using of mutagens propose the possibility of inducing desired characters that cannot be found in nature, in a variety or lost during the evolution [2].

A mutation is defined as any change within the genome of an organism, and it is not brought on by normal recombination and segregation [3]. The direct use of mutation is a very valuable supplementary approach to plant breeding. The main advantage of this technique is the shorter time required to breed a crop with improved character(s) than the hybridization process to obtain the same results.

Induced mutations consequently have a high potential for bringing about further genetic improvements. Induced mutations have played a significant role in meeting challenges related to world food and nutritional security by way of mutant germplasm enhancement and their utilization for the development of new mutant varieties. A wide range of genetic variability has been induced by mutagenic treatments for use in plant breeding and crop improvement programs [1]. Physical mutagens are generally preferred by reason of being convenient, easily reproducibility, and user-/environment-friendly method. Ionizing radiation is used as a physical mutagen in breeding applications.

2. Types of ionizing radiation

Ionizing radiation (IR) is categorized by the nature of the particles or electromagnetic waves that create the ionizing effect. These have different ionization mechanisms and may be grouped as directly or indirectly ionizing. The physical properties of ionizing radiation types, namely gamma rays X-rays, UV light, alpha-particles, beta-particles, and neutrons, are different; therefore, their potential usage and bioapplicability to the breeding programs are different.

In the beginning of the twentieth century, ionizing radiation has been begun to induce the mutations. They can be particulate or electromagnetic (EM). Their specific feature is the localized release of large amounts of energy. These have different ionization mechanisms, and they can group as directly or indirectly ionizing. The physical and chemical reactions initiate the biological effects of ionizing radiations [4].

Mostly, X-rays had been used, and later gamma rays and neutrons have been preferred. Two forms of electromagnetic radiation, X-rays or gamma (γ) rays, are widely used in biological systems and most clinical applications. Cobalt-60 and cesium-137 (Cs-137) are the main sources of gamma rays used in biological studies. Cesium-137 is more preferred since its half-life is much longer than cobalt-60. Gamma rays are produced spontaneously, whereas X-rays are produced in an X-ray tube (accelerated electrons hit a tungsten target, and then they are decelerated. The Bremsstrahlung radiation is part of the kinetic energy, belongs to the electrons, and is converted to X-rays). Energy transfer is caused by the interaction, it cannot completely displace an electron, and it produces an excited molecule/atom; whenever the energy of a particle or photon exceeds the ionization grade of a molecule, ionization occurs. Ten electronvolt binding energy for the electrons is determined for biological materials, and higher energetic photons are considered as ionizing radiation, whereas the energies between 2 and 10 eV, which cause excitation, are called as nonionizing. Electrons, protons, α-particles, neutrons, and heavy charged ions are clinically used natural radiation types [4–6].

2.1. Effects of ionizing radiations

Ionizing radiation (IR) is known to effect on plants. Their effects are classified as direct and indirect. Stimulatory, intermediate, and detrimental effects on plant growth and development are based on dose of ionizing radiation applied to the plant tissues. The main point is to evaluate the impacts of ionizing radiation at genetic level. The severity of the impacts of radiation is in relation with the species, cultivars, plant age, physiology, and morphology of the plants besides their genetic organizations.

Ionizing radiation causes structural and functional changes in DNA molecule, which have roles in cellular and systemic levels. The nature of DNA modifications includes base alterations, base substitutions, base deletions, and chromosomal aberrations. These modifications are the reasons of macroscopic phenotyping variations [5, 6].

Interaction between atoms or molecules and ionizing radiations causes free radical production that damages the cells. Free radical is defined as an atom or group of atoms including an unpaired electron. Water in the cell accumulates energy initially and facilitates the production of reactive radicals, which oxidize and reduce. They have a role in direct and indirect actions of ionizing radiations. In direct action, a secondary electron reacts directly with the target to produce an effect, while in indirect action, free radicals produced via radiolysis of water interact with the target to comprise target radicals [6, 7].

There are substantial data indicating that the lethal effects of radioactive compounds accumulate in nucleus rather than other parts. Therefore, DNA is the main target as a result of ionizing radiation, and it targets DNA directly or indirectly and leads various alterations. Direct ionization of DNA, reactions with electrons or solvated electrons, reactions with OH or H_2O^+, and reactions with other radicals can damage cellular DNA. There are some possibilities of DNA damages caused by ionizing radiation. IR and secondarily produced reactive oxygen species can cause changes in deoxyribose ring and structures of bases, DNA-DNA cross-links, and DNA-protein cross-links. Hydroxyl radicals react with bases. The reactive intermediates are produced as a result of this interaction [7, 8].

Hydroxyl radicals separate hydrogen atoms from the sugar-phosphate backbone of DNA to form 2-deoxyribose radical, which cause strong damages via attacking to oxygen or thiol groups [8]. Researchers have shown that purine and pyrimidine rings, single-strand breaks (SSBs), and base loss regions are damaged by DNA radiolysis products induced by free radicals. The amount of the yield of the individual products is important and reported to be different than produced during oxidative metabolism. Although free radicals attack on DNA and cause several DNA damages, they have not been thought to lead lethal and mutagenic results. Ionizing radiation-induced base damages are widely studied by in vitro studies. It is also reported by several studies that direct and indirect radiation effects may produce identical reactive intermediates. Oxygen is another key molecule that determines the biological effectiveness of the ionizing radiation. Oxygen can easily react with many free radicals. The amount of the radicals presents in deoxyribose or bases; harmful DNA damages occur [9–11].

If the damage site is deoxyribose, a strand brake directly forms. DNA base damages like ring saturation destabilize the N-glycosidic bonds, and abasic deoxyribose residues form. These regions can be converted into strand breaks. Double-strand breaks (DSBs) happen as a result of a localized attack by two or more OH radicals on DNA. Another potential reason can be defined as a hybrid attack that OH damages one of the strands, whereas the other strand exposes to a direct damage within 10 base pairs of the hydroxyl radical [12]. IR leads chromosomal aberrations during cell division. Chromosome malsegregation and defects in chromatid separation, bridge formation, chromosome exchange, chromosome breakage, and loss of chromosome fragments can be observed after IR treatment [13].

We can classify IR as an abiotic stress factor; therefore, the plants represent different levels of adaptive responses. DNA repair mechanisms and adaptive responses against radiation could protect the plant genome from excessive modifications. Natural ionizing radiation is supposed to play a significant role in the evolution of the plants. Homolog recombinations between the chromosomes would result in formation of new altered generations that show specific adaptive capabilities [13].

3. Mutation breeding

In nature, mutations acquired new survival traits to the crops against environmental stresses both biotic and abiotic. Many of these survival traits could be weakened or totally lost in time. Mutations are sudden changes at the genotype level and cause small and exquisite changes in phenotype, which cannot be detected by advanced molecular techniques. Identification of naturally mutated gene is inconvenient. When the breeders pinpoint the mutated gene, wild-type features have to be reestablished. This task is becoming increasingly infeasible due to long time, more human source, and increase in cost. That's why new breeding strategies were needed to be improved to fortify the crops. To achieve this mission, plant breeders should rebuild in crop plants several specific traits, which have role in survival of the plants under extreme conditions providing the other crop-specific traits such as quality, yield, etc. Phenotyping-based processes of conventional breeding strategies should have moved from base to a high level of genotype-based breeding methods [1, 14]. New technologies should be legal, economic, and ethical for the breeders and the consumers.

Under such circumstances, inducing mutations are potential applications to produce crops with desired traits and easily selected from the germplasm pool. As described above, radiation can cause several effects on genetic material due to the exposure dose. These effects can be classified in both positive and negative approaches. Beside the detrimental effects of radiation, plant breeders are focused on the effective usage of gamma radiation in breeding programs. Changes in agronomic characters can be transmitted to the next generations. Nuclear techniques are begun to be used in plant breeding mostly for inducing mutations. During the past 60 years, we observed a significant increase in the major crops. Ionizing radiations such as X-rays and gamma rays have been used for improvement of several crops such as wheat, rice, barley, cotton, tobacco, beans, etc. [15]. Plant breeders are also combined with this resource with different techniques to increase the efficiency and shorten the time. Induced mutagenesis and combined breeding strategies are effective to improve quantitative and qualitative traits in crops in a much shorter time than the conventional breeding procedure [6].

Gamma radiation is widely used to induce mutations in breeding studies than chemical mutagens. Ionizing radiation could cause several DNA damages randomly; therefore, several mutations (from point mutation to chromosome aberrations) could be induced. Over 3000 mutant varieties of major crops have been reported to be developed by ionizing radiation [2, 16].

Mutation rate/mutation frequency is defined as the ratio of mutation per locus and also termed as the number of mutant plant per M_2 generation [16]. It changes due to per dose and mutagen. The main point is to determine the best dose for inducing mutants rather than its type. From past to present, it is concluded that the doses between LD50 and LD30 (doses lead to 50% and 30% lethality) are generally useful in mutation breeding programs. The importance of convenient dose that depends on the radiation intensity and exposure time is gestured by the researchers [6].

The final target is to select the desired mutants in the second and third generations (M_2 and M_3). It is effective to select the mutants treated by the mutagens with a high mutation frequency from the M_1 population. M_1 population consists of heterozygous plants. That means during the treatment one allele is affected by the mutation, and it is impossible to discriminate the recessive mutation in this generation. Therefore, the breeders should sift out the next generations to identify the homozygotes for both dominant and recessive alleles [6]. M_2 population is the first generation that the selection begins. Physical, mechanical, phenotypic, and other methods are used for the selection of the mutants. When the plant breeder finds a mutant line, the next step is the multiplication of the seeds for further field and other studies. The main theme is to develop a mutant, which has a potential to be commercial variety surpassing the mother cultivar or a new genetic stock having improved properties.

According the 2015 data of Food and Agriculture Organization/International Atomic Energy Agency (FAO/IAEA), over 232 different crops including wheat, rice, sunflower, soybean, tomato, and tobacco were subjected to mutation breeding programs and over 3000 mutant varieties with improved properties in over 70 countries [6]. The mutant plant production distribution worldwide is given in **Figure 1**.

Sixty-one percent of these varieties was improved by using gamma radiation. **Figure 2** represents the maximum plant species improved via mutation breeding.

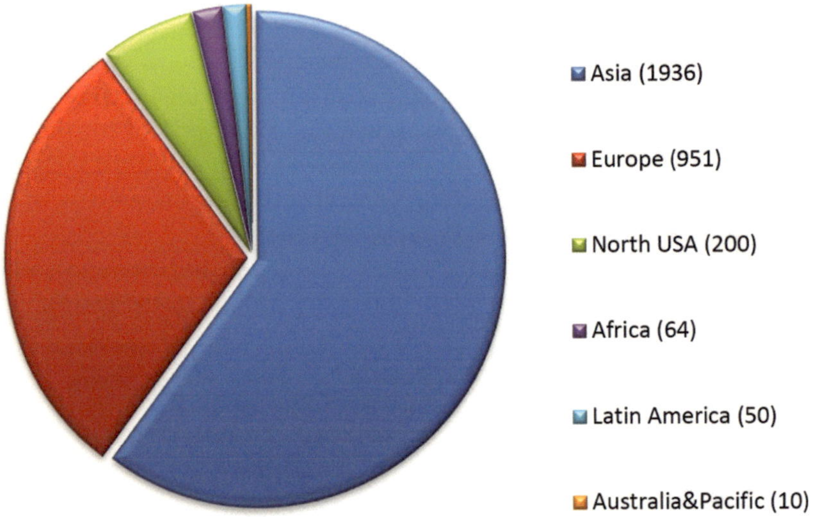

Figure 1. The number and the rate of the mutant cultivar production rate worldwide [17].

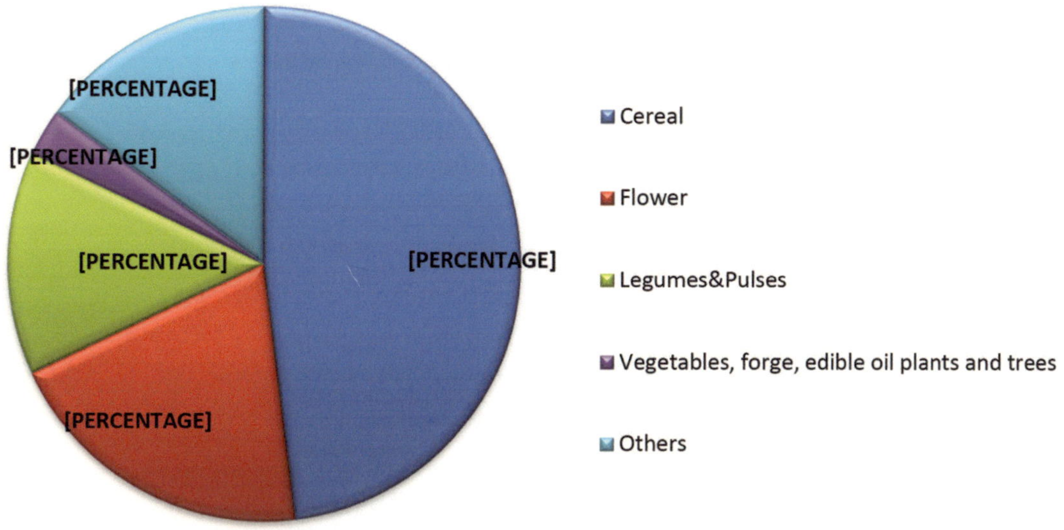

Figure 2. The maximum plant species improved via mutation breeding [17].

Mutation breeding studies are widely preferred to improve cultivars tolerant or resistant to various abiotic stresses and biotic stresses such as bacteria, viruses, and pathogens and to improve the quality and the agricultural traits of the crops such as oil, protein, and yield [6]. The most improved features in some plant species by gamma radiation were given in **Table 1**.

Instead of waiting for natural mutations to generate a desired trait, creating a mutation with different tools may promote to the breeding studies. The simplicity and low cost of mutation treatments and gamma radiation became an effective tool to improve new agronomic traits in various crops. It may be evaluated as an alternative to genetically modified plants. The released mutation breeding-derived varieties showed the potential usage of mutation breeding as a flexible and available accession to any crop supplied for desired purposes, and discriminating techniques are successfully combined.

As mentioned above, mutation breeding studies are provided to numerous researches in terms of developing applications for plant biotechnology, plant tissue culture, and mutation treatments to improve new cultivars. Therefore, research and developmental studies are widely associated to combined techniques including in vitro culture and molecular techniques through mutation breeding. In vitro mutagenesis applications are becoming important at this point.

3.1. In vitro mutagenesis applications

Induced mutagenesis is a widely used method to identify and isolate the plant genes in combination with molecular accessions. These kinds of studies supply a clear prehension into the relation of genes and functions of the genes that have role in growth and development under several conditions [12].

Crops	Improved traits
Apple	Early maturing, red fruit skin color, variegated leaf, dwarf, compact tree, resistance to powdery mildew and apple scab
Apricot	Earliness
Banana	Earliness, bunch size, reduced height, tolerance to *Fusarium oxysporum* f. sp. cubense, large fruit size, putative mutant resistant to black sigatoka disease
Barley	Phytate (antinutrient)
Canola	Oil quality improvement
Chickpea	Resistant to Ascochyta blight and Fusarium wilt
Citrus	Seedless, red color fruit, *Xanthomonas citri* disease resistant, resistant to tristeza virus
Cotton	Resistant to bacterial blight, cotton leaf curl virus, high fiber
Indian jujube	Fruit morphology, earliness
Loquat	Fruit size
Maize	Resistant to pathogen *Striga*, acidity, and drought tolerance; improvement of protein quality
Mung bean	Resistant to yellow mosaic virus
Pineapple	Spineless, drought tolerance
Tomato	Resistant to bacterial wilt (*Ralstonia solanacearum*)
Rapeseed	Resistant to powdery mild
Rice	Resistant to blast, yellow mottle virus, bacterial leaf blight and bacterial leaf stripe, semidwarf/dwarf cultivar, lodging resistance, acid sulfate soil tolerance; tolerant to cold and high altitudes, salinity tolerance, early maturity, high-resistant starch in rice for diabetes patients, giant embryos of eight more plant oil, low amylose, low protein)
Soybean	Resistant to Myrothecium leaf spot and yellow mosaic virus, oil quality improvement, oilseed meals that are low in phytic acid desirability, poultry and swine feed
Strawberry	Thick and small leaf, light leaf color, white flesh and long fruit, *Phytophthora cactorum* resistance
	Salinity tolerance
Sunflower	Oil quality improvement, semidwarf/dwarf cultivars
Wheat	Resistant to stripe rust

Table 1. Applications of induced mutagenesis for improved features in plant breeding [6, 17].

In vitro culture methods appear to have opportunities to display the useful variants. The recent improvements on in vitro technology acquired an importance to enlarge the aim of mutation breeding applications. The use in conjunction of in vitro tissue culture and mutation breeding methods makes a significant contribution to improve new crops and new varieties. Jain [18] reported the importance of this technique for ornamental plants beside the crops.

It is known that genetic variabilities may occur during in vitro culture conditions without any application of mutagens, spontaneously. The frequency of the variants still indeterminable, and there are many parameters to depend on. Application of the mutagens can increase the rate of genetic variability via inducing the frequency of the mutations.

The progress in recombinant DNA technologies and genes can be easily cloned from a genome into a genome of an organism. Genes can be purified in vitro in small amounts, and therefore the potential of inducing mutations has significantly broadened. In a controlled experimental environment, it is available to change the sequence of the nucleotides of DNA. In vitro mutagenesis studies systemically and efficiently focus on the potential ways of inducing mutagenesis. Some applications of mutagenesis depend on using isolated DNA molecule. In contrast to conventional mutagenesis, in vitro mutation breeding can be thought as a practicable and achievable technique to improve new genetic variabilities. Only few traits can be modified, and the remaining is not altered by the treatment.

In vitro mutagenesis have some properties such as increased mutation rate, uniform mutagen treatment, needs of less space and time for large populations, and opportunity to keep the plant material disease-free. On the other hand, one of the main restrictions of mutation breeding application is the formation of chimeras as a result of the treatment. At this point, mutant selection process is becoming important.

In vitro culture methods are more useful in mutation studies. Totipotency is a natural feature of the plant cells. By using one plant cell, it is possible to produce a whole plant, induce regeneration of the tissues and micropropagation of the plants in large volumes, and give opportunity to use different parts of the plants (stem, leaves, cuttings, apical and axillary buds, and tubers) to induce mutagenesis easily. Another advantage of in vitro culture is to screen the populations after mutagenesis to select the variant/mutants before giving a whole plant. Different plant tissues can be propagated to produce different tissues by using several combinations of plant growth regulators. Callus is an important cell organization. Cell suspension culture technique is started by using callus tissue to separate the single totipotent cells. Every plant cell can be differentiated into somatic embryos which is a useful tool for mutagenesis [4, 9].

The target of the studies is to isolate the non-chimeric mutants from the irradiated explants to obtain desired mutants via repetitive selection processes. Meanwhile, duration of the culture and the selective traits that mutated are the main factors effect these processes [4]. M_2 generation of the culture is the earliest step that the predominantly recessive mutants could be determined. Mutagen treatment can be applied at different stages of the cultures.

3.1.1. *Selection of species and explant types for in vitro culture*

Correct choice of the plant species due to economic, commercial production capacity and agricultural importance is the first step of an in vitro mutagenesis study. The selection of the plant material is related to the success of the in vitro culture. Seed, callus, node, shoot, and root tip cultures are the most commonly preferred plant material for in vitro mutagenesis applications. The genotype of a plant has a role in in vitro culture studies. The studies showed that different explants of same plant had different responses to the same radiation dose [19, 20]. Therefore, it is necessary to design an in vitro mutagenesis experiment in a proper combination of dose and explant type.

3.1.2. Determination of proper gamma radiation dose

The most important subject of in vitro mutagenesis is to select the suitable radiation dose to obtain the maximum viability. In the beginning, assessment of the LD_{50} value is needed to optimize the exact mutation dose. The sensitivity of the plants changes due to the species, cultivars, and current physiological environment. A preliminary dose experiment should be performed to define the appropriate dose. Reduced growth and seedling damages may be seen as traces of the genetically damaged plants after irradiation [21]. IAEA/FAO reported the average doses for crop species and summarized in **Table 2**.

According to the findings of the preliminary studies done with gamma radiation, it has been reported that there is no linearity between the radiation dose and the variance. The experimental gamma radiation treatments were summarized in **Table 3**.

Seed, callus, shoot tips, node cultures, and bulblets were frequently used for irradiation of different species. ^{137}Cs and ^{60}Co gamma sources were used to induce mutagenesis at different doses depending on the radiosensitivity of the explants. Atak et al. [24] used 100–500 Gy radiation doses produced by ^{137}Cs gamma source for soybean seeds, while Singh and Datta [29] used ^{60}Co gamma source at different doses ranging between 10 and 100 Gy for *Triticum aestivum* seeds. Ulukapı et al. [21] also used ^{60}Co gamma source at 80–240 Gy radiation doses to induce genetic variability for *Solanum melongena* L. Çelik and Atak [23] used 100, 200, 300, and 400 Gy gamma rays by ^{137}Cs to determine the effective radiation dose for breeding studies of two Turkish tobacco varieties. They irradiated the tobacco seeds and selected the salt-tolerant mutants in M_3 progeny.

Seetohul et al. [35] used 0–60 Gy gamma doses of ^{60}Co gamma source to induce mutations for shoot tip explants of Taro plant. Jain [26] irradiated shoot tip explants of *Musa* spp. by Cesium-137 at 10–50 Gy doses, while Baraka and El-Sammak [33] used 0.25–1 Gy for *Gypsophila paniculata* L. shoot tip explants by ^{60}Co gamma source. Atak et al. [25] used shoot tip explants of *Rhododendron* varieties to induce mutants at 5–50 Gy of gamma rays of ^{137}Cs source.

In tissue culture treatments, different synthetic chemicals show similar effects as plant growth regulators which have abilities to induce growth of the tissues as desired. In mutation-based

Species	Useful mutation breeding dose (gray)
Oryza sativa japonica	120–250
Oryza sativa indica	150–250
Triticum aestivum	40–70
Hordeum vulgare	30–60
Glycine max	100–200
Phaseolus vulgaris	80–150
Nicotiana tabacum	200–350
Medicago sativa	400–600

Table 2. Gamma radiation radiosensitivity of some crop species [22].

Species	Plant material	Gamma ray source	Gamma radiation dose (Gy)	Reference
Nicotiana tabacum L.	Seeds	^{137}Cs	100–400	[23]
Glycine max L. Merr.	Seeds	^{137}Cs	100–500	[24]
Rhododendron spp.	Shoot tip	^{137}Cs	5–50	[25]
Musa spp.	Shoot tip	^{137}Cs	10–50	[26]
Solanum tuberosum L. "Marfona"	Node	^{137}Cs	5–50	[27]
Saccharum officinarum L.	Leaf primordia	^{137}Cs	10–50	[28]
Triticum aestivum	Seed	^{60}Co	10–100	[29]
Paphiopedilum delenatii	PLBs, shoot buds, in vitro plantlets	^{60}Co	10–80	[30]
	Embryogenic callus	^{60}Co	50–400	[31]
Saccharum officinarum L.	Embryonic callus culture	^{60}Co	10–80	[32]
Solanum melongena L.	Seed	^{60}Co	80–240	[21]
Gypsophila paniculata L.	Shoot tips and lateral buds	^{60}Co	0.25–1	[33]
Etlingera elatior	Axenic culture	^{60}Co	10–140	[34]
Colocasia esculenta L. Schott	Shoot tips	^{135}Co	0–60	[35]
Chrysanthemum grandiflora	Shoots	^{60}Co	5–30	[36]
Chrysanthemum morifolium	Ray florets	^{60}Co	0.5–1	[37]
Lilium longiflorum Thunb. Cv. White fox	Bulblets	^{60}Co	0.5–2.5	[38]
Rosa hybrida L.	Single-node cuttings	^{60}Co	5–80	[39]

Table 3. In vitro mutagenesis protocols for some crop species.

selection of the plants with desired characters using in vitro cultivation methods for vegetative plants, clonally reproduction of the plant parts is needed in order to detect the mutant generations via using easy stability tests [27]. The schematic diagram representing the usage of gamma radiation for in vitro mutagenesis applications is given in **Figure 3**.

3.1.3. In vitro selection of the mutants

The selection of the desired mutants is an essential and important part in a mutation breeding program. In vitro mutagenesis applications give opportunity to the breeders to select the mutants in a controlled environment. The plant breeders can work with a large population of plant material. Different culture techniques such as suspension cultures and protoplast cultures can be widely preferred to have a genetic uniformity in the selection studies.

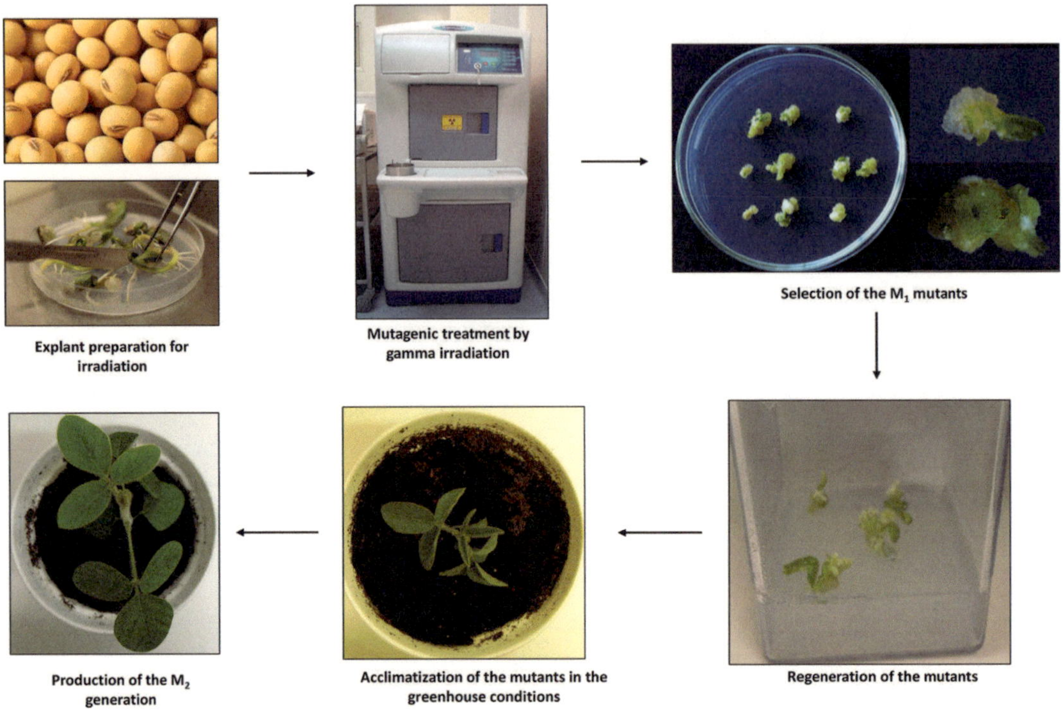

Figure 3. The representative schematic diagram of an in vitro mutagenesis application.

In vitro selection studies have some advantages. These can be classified as follows:

1. Easiness of the application
2. Reduced time of the selection
3. Availability to use some selective agents in culture conditions

In vitro selection studies can be performed in two types: single step and multistep [4]. In single-step selection procedure, the inhibitor agent is added to the culture environment and the subcultures used for the selection studies. In multistep method, the dose of the selective agent below lethal dose is added to the culture, and the concentration of the inhibitor is gradually increased in subcultures. The selected mutant by this method has been defined as more stable than selected via other methods [4].

Food and ornamental plants are widely assessed for nutritional quality, early ripening, better flower, and biotic/abiotic stress tolerance capacities [4]. For abiotic stress treatments, it is more convenient to control the culture conditions than in the field environment [20, 40]. Salt, drought, cold, and heavy metal tolerance have been successfully performed in many plants. Callus, suspension cultures, or protoplast cultures were used for in vitro selection analysis by adding the selective agents reducing the growth such as mannitol and polyethylene glycol for drought tolerance; NaCl for salt tolerance; boron, aluminum, and nickel for metal tolerance; or changing the temperature of the cultures to select cold/high-temperature-tolerant plants [4, 41]. Both selection strategies, single step and multistep, can be used. The main point is to inhibit the false-positive selection responses due to epigenetic alterations in long-term culture conditions. When the plants are subject to long-term stress treatments with gradual increase of the selective agent, non-tolerant cells can experience stable epigenetic alterations, which can be inherited by mitosis. In order to avoid this period, preference of single-step selection procedure is suggested to be efficient during mutation breeding programs [41].

In selection studies, the main criterion is to define the exact selective agent. This means that the molecular mechanism of the desired trait should be clearly understood by the plant breeders. Morphological and physiological changes should be used in combination to discriminate the mutants. All the parameters such as leaf injury, slower growth, average number of shoots per explant, survival percentage of the plants, fresh weight of the explants, leaf photosynthetic capacity, antioxidant defense system, and accumulation of osmolytes should be investigated in detail especially for stress tolerance studies [40, 41].

3.2. Mutational genomic analysis

Mutational genomics is becoming a valuable tool to differentiate the mutants improved via mutation breeding programs. It is also an important tool to understand the molecular basis of the plant stress response based on the data gathered from mutants of model plants and an easy way to determine the genetic similarities and characterize the variations between the mutants at the DNA level.

The mutants were identified based on morphological characters, traditionally. The new developments in DNA technologies give opportunity to the plant breeders to make it quick and definite.

Molecular markers are widely used to differentiate the genetic differences between the mutant and the mother plants through characterizing the variations at DNA level. High-throughput genomic platforms such as random amplified DNA polymorphism (RAPD), cDNA-amplified fragment length polymorphism (AFLP), single-strand conformational polymorphism (SSCP), microarray, differential display, targeting induced local lesions in genome (TILLING) and high-resolution melt (HRM) analysis allow rapid and in-depth global analysis of mutational variations [4].

Among these methods RAPD, inter simple sequence repeat (ISSR), and AFLP have been frequently used in genomic classification of the mutants [15, 42]. RAPD is an inexpressive and a rapid method to use in many fields of biotechnology. There is no need for genome information. It has been widely used to determine the genetic diversity in mutation breeding programs of many plants. RAPD is an efficient method to detect DNA alteration via using random primers. It has been started to use in earlier studies of genetic variabilities obtained by radiation treatments in *Chrysanthemum* [36, 37], soybean [24], sugarcane, sunflower, groundnut [43], tobacco [23], potato [27], *Rhododendron* [25]. ISSR method is another molecular marker method widely used in plant biotechnology applications. It is also easy to apply more informative than RAPD, reliable, and inexpensive [44, 45]. ISSR primers are designed by using microsatellite sequences to amplify the genomic regions flanked by microsatellite repeats. By using one primer, it is possible to amplify multiple fragments as a result of ISSR analysis [46, 47]. The information obtained from ISSR analysis is more reliable than RAPD to provide supplementary data of the genetic variations of the mutants from the nonoverlapping genome regions [48].

Xi et al. [38] reported an in vitro mutagenesis protocol for *Lilium longiflorum* Thunb. cv. White fox. They used 0, 0.5, 1.0, 1.5, 2.0, and 2.5 Gy gamma rays to observe the effects of radiation on adventitious bud formation from bulblet-scale thin cell layers. 1.0 Gy was determined as the most effective dose due to survival rate of the bulblet-scale thin layers. They also evaluated the morphological mutants using ISSR DNA fingerprinting method.

Sianipar et al. [49] used RAPD method to detect the genetic variability between the mutant plantlets improved from gamma-irradiated rodent tuber calli. They obtained 69 fragments from 11 mutant plantlets by using 10 RAPD primers.

Barakat and El-Sammak [33] irradiated shoot tips and lateral buds of *G. paniculata* with four different gamma radiation doses between 0.25 and 1 Gy. They detected the genetic polymorphisms among the mutants by RAPD analysis. They obtained 105 different amplification products from 10 random primers. RAPD is evaluated as an efficient molecular marker technique to detect the variations. Atak et al. [25] used RAPD method to show the genetic similarities of the *Rhododendron* mutants improved via gamma irradiation. They used 0–50 Gy gamma radiation doses to improve the shoot and root regeneration rates of *Rhododendron* plants. RAPD detected higher genetic variability among the *Rhododendron* mutants. Yaycılı and Alikamanoğlu [27] observed 89.66% polymorphism rate with six primers among the mutant potato plants, which were improved as salt tolerant via gamma radiation treatment. Kaul et al. [36] used in vitro mutagenesis in *Chrysanthemum* cv. Snow Ball by irradiation of the in vitro shoots, and genetic polymorphisms among the mutants and the control plants were assessed by RAPD. They reported that 10 Gy gamma irradiation was found as the most effective dose to induce genetic variation in morphological traits, and they observed 100% polymorphism among the mutants. Gamma radiation-induced salt-tolerant oriental tobacco mutants were improved by Çelik and Atak [23]. Salt tolerance of the mutants was controlled by the callus induction in the presence of high salt concentration. The genetic similarities of the mutants were determined by RAPD analysis. The relationships between the salt-tolerant mutants and controlled tobacco varieties were shown in Unweighted Pair Group Method with Arithmetic Mean (UPGMA) dendrogram. Some representative RAPD profiles of the mutants developed by in vitro mutagenesis were given in **Figure 4**.

Sen and Alikamanoğlu [44] used ISSR method to differentiate the drought-tolerant sugar beet mutant improved via irradiation of shoot tip explants by gamma radiation. They obtained 91 polymorphic bands of 106 PCR fragments with 19 inter simple sequence repeat (ISSR) primers.

Perera et al. [40] applied in vitro mutagenesis treatment to an important energy crop giant miscanthus (*Mischanthus × giganteus*) to induce variation in cultivar Freedom. ISSR markers were used to determine the variations in the mutant plants. The putative mutants were selected due to the results of molecular marker analysis to use for further bioenergy researches. Wu et al. [50] used ISSR analysis to show the genetic similarities between the mutants. For this reason, they used 60 ISSR primers, and 60 polymorphic bands of 392 were evaluated to have information on the molecular level of mutation breeding. Atak et al. (unpublished data) [51] used ISSR marker method (with 61 ISSR primers) to define the genetic variation among the 8 salt-tolerant mutant soybeans obtained from in vitro mutagenesis treatment by using ^{137}Cs gamma source. The representative results of ISSR amplification of 8 salt-tolerant mutant soybeans were given in **Figure 5**.

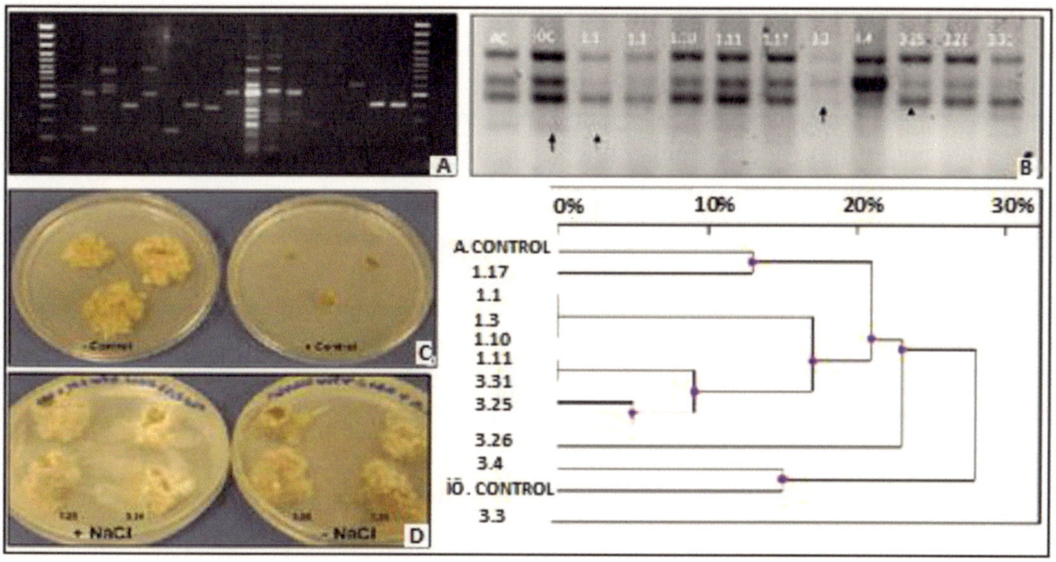

Figure 4. Evaluation of salt-tolerant tobacco mutants improved via in vitro mutagenesis application. A and B represent the RAPD profiles of the mutants. C shows the callus growth of control plants under in vitro salt stress. D shows the callus growth profiles of the mutants under salt stress [23].

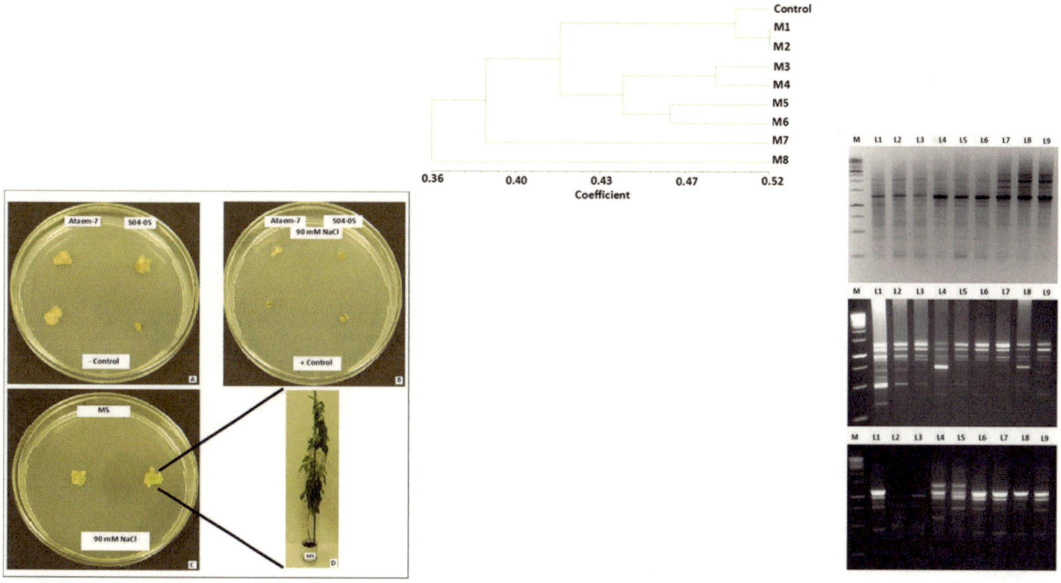

Figure 5. The representative results of evaluation of 8 salt-tolerant mutant soybean plants improved by in vitro mutagenesis treatment. A.The callus gowth of Ataem-7 and S04-05 soybean cultivars. B. The callus gowth of Ataem-7 and S04-05 soybean cultivars under 90 mM NaCl. C. Callus growth of M5 under 90 mM NaCl D. The whole plant M5.

Single-strand conformational polymorphism (SSCP) is another strength method to identify the variations between the mutant and mother plants in amplified DNA samples. It is widely used to determine the genetic mutations in several organisms. It is also an effective method to find a potential genetic marker which is in relation with a desired trait to use in selection studies of agricultural populations [52]. Irradiation of the plant tissues can cause mutation between the allelic gene copies [single-nucleotide polymorphism (SNP)]. SSCP is an efficient method to detect these polymorphisms. It is possible to detect relations between SSCP polymorphisms and quantitative traits [53].

These methods can only be able to detect the genetic variations of the mutants in accordance with the mother plants. There are a number of methods to screen the causal mutation at a desirable phenotype. Molecular markers that are in relation with the mutation are known to be able to segregate in the next progenies. The main point is to make the functional analysis of the mutant genes that have role in acquiring the new desired characters. To identify a mutant, the number of the genes controlling that specific phenotypic character is deterministic [54].

In a mutation breeding program, identification of differentially expressed genes, the biological processes they have role in, or the metabolic pathways of interest should be carried out through modern genomics and system biology. To achieve this, there are specific tools to discriminate with the use of next-generation molecular techniques. In microarray systems, it is available to detect the gene expressional differences between the mutants and control plants. Thousands of spots on a microarray chip containing a few million copies of identical DNA molecules buried on each spot are related to each gene of a plant genome. If it is a targeted mutation, it is possible to show the expressional differences between them by microarray technique. In general, spontaneous mutations cannot be detected at microarray systems. Sequencing methods are more efficient in the meanwhile. Mutant plants can now easily sequence by next-generation sequencing (NGS) techniques to define the mutations [55]. To apply these methods, there is no need for a reference genome. These analyses can be classified as forward genetic screening methods that give opportunity to improve the knowledge about the genes that control specific biological roles in mutant plants. In contrast to forward genetic, reverse genetic is more popular to detect the function of a gene. In mutation breeding programs, the plant breeders are focused to identify the individuals from a population that have an allelic variation of a gene. As mentioned previously, these individuals are improved by mutagenic treatments. TILLING method is available to determine the mutants with specific phenotypes. In tomato, approximately 3000 mutant lines that were improved by chemical mutagens on fruit ripening trait were identified by this method. This method is used for barley to screen the homeodomain-leucine zipper protein mutants. Recent progresses in NGS technologies and TILLING which is in relation with these technologies make it possible to screen the potential genes [54, 56].

4. Discussion and the conclusion

The increasing importance of plant breeding studies in correlation with biotechnology and molecular genetics is attempted to meet the requirements of increasing population for food and crop plants. Therefore, mutation breeding treatments have become more frequent and alternative to classical breeding and genetically modified plants. The main aim is to combine several features of many plants in one super plant. In vitro mutagenesis has become an efficient tool for this purpose. Plant breeders are focused to crop improvement techniques to improve genetic variations of useful traits by using next-generation molecular methods.

Using these advanced genomic techniques, new molecular mechanisms and new genes can be potentially identified by the plant breeders as a result of in vitro mutagenesis treatments. To gain more data, additional needs of various comparative and descriptive experiments can be upgraded to acquire more specific points to build the relations between the regulatory mechanisms. Therefore, the recent progress in mutation breeding studies in relation with new technologies is quite important to contribute new advancement to plant breeding programs.

Author details

Özge Çelik* and Çimen Atak

*Address all correspondence to: ocelik@iku.edu.tr

Department of Molecular Biology and Genetics, Faculty of Science and Letters, TC İstanbul Kultur University, Istanbul, Turkey

References

[1] Schaart JG, van de Wiel CCM, Lotz LAP, Smulders MJM. Opportunities for products of new plant breeding techniques. Trends in Plant Science. 2016;**21**:438–448. DOI: 10.1016/j.tplants.2015.11.006

[2] Pathirana R. Plant mutation breeding in agriculture. CAB Reviews: Perspectives in Agriculture, Veterinary Science, Nutrition and Natural Resources. 2011;**6**:107–126. DOI: 10.1079/PAVSNNR20116032

[3] Phillip RL, Rines HW. Expanding the boundaries of gene variation for crop improvement. In: Shu QY, editor. Induced Plant Mutations in Genomics Era. Rome: FAO&IAEA; 2009. pp. 21–26.

[4] Penna S, Vitthal SB, Yadav PV. In vitro mutagenesis and selection in plant tissue cultures and their prospects for crop improvement. Bioremediation, Biodiversity, Bioavailability. 2012;**6**:6–14.

[5] Predieri S. Mutation induction and tissue culture in improving fruits. Plant Cell, Tissue and Organ Culture. 2001;**64**:185–210.

[6] Oladosu Y, Rafii MY, Abdullah N, Hussin G, Ramli A, Rahim HA, Miah G, Usman M. Principle and application of plant mutagenesis in crop improvement: A review. Biotechnology&Biotechnological Equipment. 2015;**30**:1–16. DOI: 10.1080/13102818.2015.1087333

[7] Limoli CL, Kaplan MI, Giedzinski E, Morgan WF. Attenuation of radiation-induced genomic instability by free radical scavengers and cellular proliferation. Free Radical Biology & Medicine. 2001;**31**:10–19. DOI: 10.1016/S0891-5849(01)00542-1

[8] Hosseinimehr SJ. Flavonoids and genomic instability induced by ionizing radiation. Drug Discovery Today. 2010;**15**:907–918. DOI: 10.1016/j.drudis.2010.09.005

[9] De Micco V, Arena C, Pignalosa C, Durante M. Effects of sparsely and densely ionizing radiation on plants. Radiation Environmental Biophysics. 2011;**50**:1–19. DOI: 10.1007/s00411-010-0343-8

[10] Wi SG, Chung BY, Kim JS, Kim JH, Baek MH, Lee JW, Kim YS. Effects of gamma irradiation on morphological changes and biological responses in plants. Micron. 2007;**38**:553–564. DOI: 10.1016/j.micron.2006.11.002

[11] Ballarini F, Carante MP. Chromosome aberrations and cell death by ionizing radiation: Evolution of a biophysical model. Radiation Physics and Chemistry. 2016;128:18–25. DOI:10.1016/j.radphyschem.2016.06.0099

[12] Shirley BW, Hanley S, Goodman HM. Effects of ionizing radiation on a plant genome: Analysis of two Arabidopsis transparent testa mutations. The Plant Cell. 1992;**4**:333–347. DOI: 10.1105/tpc.4.3.333

[13] Esnault MA, Legue F, Chenal C. Ionizing radiation: Advances in plant response. Environmental and Experimental Botany. 2010;**68**:231–237. DOI: 10.1016/j.envexpbot.2010.01.007

[14] Barabaschi D, Tondelli A, Desidero F, Volante A, Vaccino P, Vale G, Cattivelli L. Next generation breeding. Plant Science. 2016;**242**:3–13. DOI: 10.1016/j.plantsci.2015.07.010

[15] Ahloowalia BS, Maluszynski M. Induced mutations – A new paradigm in plant breeding. Euphytica. 2001;**118**:167–173. DOI: 10.1023/A:1004162323428

[16] Tanaka A, Shikazono N, Hase Y. Studies on biological effects of ion beams on lethality, molecular nature of mutation, mutation rate, and spectrum of mutation phenotype for mutation breeding in higher plants. Journal of Radiation Research. 2010;**51**:223–233. DOI: 10.1269/jrr.09143

[17] Ulukapı K, Nasırcılar AG. Developments of gamma ray application on mutation breeding studies in recent years. In: International Conference on Advances in Agricultural, Biological & Environmental Sciences; 22–23 July; London. UK; 2015. pp. 31–34. DOI: 10.15242/IICBE.C0715244

[18] Jain SM. Mutation-assisted breeding for improving ornamental plants. In: Mercuri A, Schiva T, editors. XXII International Eucarpia Symposium, Section Ornamentals, Breeding for Beauty; 1 September; San Remo. Italy: ISHS Acta Horticulturae; 2006. DOI: 10.17660/ActaHortic.2006.714.10

[19] Zhou LB, Li WJ, Ma S, Dong XC, Yu LX, Li Q, Zhou GM, Gao QX. Effects of ion beam irradiation on adventitious shoot regeneration from in vitro leaf explants of *Saintpaulia ionahta*. Nuclear Instruments and Methods in Physics Research B. 2006;**244**:349–353. DOI: 10.1016/j.nimb.2005.10.034

[20] Gallone A, Hunter A, Douglas GC. Radiosensitivity of Hebe 'Oratia Beauty' and 'Wiri Mist' irradiated in vitro with γ-rays from ^{60}Co. Scientia Horticulturae. 2012;**138**:36–42. DOI: 10.1016/j.scienta.2012.02.006

[21] Ulukapı K, Özdemir B, Onus AN. Determination of proper gamma radiation dose in mutation breeding in eggplant (*Solanum melongena* L.). In: Mastorakis NE, editor. Proceedings of the 4th International Conference on Agricultural Science, Biotechnology, Food and Animal Science (ABIFA'15); 22–24 February; Dubai. United Arab Emirates: WSEAS Press; 2015. p. 149.

[22] Conger BV, Konzak CF, Nilan RA. Mutagenic radiation. In: Manual on Mutation Breeding. 2nd ed. Vienna: FAO/IAEA; 1977. pp. 40–50.

[23] Çelik Ö, Atak Ç. Random amplified polymorphic DNA analysis of salt-tolerant tobacco mutants generated by gamma radiation. Genetics and Molecular Research. 2015;**14**(1): 1324–1337. DOI: 10.4238/2015.February.13.12

[24] Atak Ç, Alikamanoğlu S, Açık L, Canbolat Y. Induced of plastid mutations in soybean plant (*Glycine max* L. Merrill) with gamma radiation and determination with RAPD. Mutation Research. 2004;**556**:35–44. DOI: 10.1016/j.mrfmmm.2004.06.037

[25] Atak Ç, Çelik Ö, Açık L. Genetic analysis of Rhododendron mutants using random amplified polymorphic DNA (RAPD). Pakistan Journal of Botany. 2011;**43**(2):1173–1182.

[26] Jain SM. Mutagenesis in crop improvement under the climate change. Romanian Biotechnological Letters. 2010;**15**:88–106.

[27] Yaycılı O, Alikamanoglu S. Induction of salt-tolerant potato (*Solanum tuberosum* L.) mutants with gamma irradiation and characterization of genetic variations via RAPD-PCR. Turkish Journal of Biology. 2012;**36**:405–412. DOI: 10.3906/biy-1110-14

[28] Khan IA, Dahot MU, Seema N, Yasmin S, Bibi S, Raza S, Khatri A. Genetic variability in sugarcane plantlets developed through in vitro mutagenesis. Pakistan Journal of Botany. 2009;**41**:153–166.

[29] Singh B, Datta PS. Gamma irradiation to improve plant vigour, grain development, and yield attributes of wheat. Radiation Physics and Chemistry. 2010;**79**:139–143. DOI: 10.1016/j.radphyschem.2009.05.025

[30] Luan LQ, Uyen NHP, Ha VTT. In vitro mutation breeding of Paphiopedilum by ionizing radiation. Scientia Horticulturae. 2012;**144**:1–9. DOI: 10.1016/jscienta.2012.06.028

[31] Khan SJ, Khan UH, Khan RD, Iqbal MM, Zafar Y. Development of sugarcane mutants through in vitro mutagenesis. Pakistan Journal of Biological Sciences. 2000;**3**:1123–1125.

[32] Nikam AA, Devarumath RM, Ahuja A, Babu H, Shitole MG, Suprasanna P. Radiation-induced in vitro mutagenesis system for salt tolerance and other agronomic characters in sugarcane (*Saccharum officinarum* L.). The Crop Journal. 2015;**3**:46–56. DOI: 10.1016/j.cj.2014.09.002

[33] Barakat MN, El-Sammak H. In vitro mutagenesis, plant regeneration and characterization of mutants via RAPD analysis in Baby's breath *Gypsophila paniculata* L. Australian Journal of Crop Science. 2011;**5**:214–222.

[34] Yunus MF, Aziz MA, Kadir MA, Daud SK, Rashid AA. In vitro mutagenesis of *Etlingera elatior* (Jack) and early detection of mutation using RAPD markers. Turkish Journal of Biology. 2013;**37**:716–725. DOI: 10.3906/biy-1303-19

[35] Seetohul S, Puchooa D, Ranghoo-Sanmukhiya VM. Genetic improvement of Taro (Colocasia esculenta var esculenta) through in-vitro mutagenesis. Uom Research Journal. 2008;**13A**:79–89.

[36] Kaul A, Kumar S, Ghani M. In vitro mutagenesis and detection of variability among radiomutants of chrysanthemum using RAPD. Advances in Horticultural Sciences. 2011;**25**:106–111. DOI: 10.13128/ahs-12775

[37] Barakat MN, Fattah RSA, Badr M, El-Torky MG. In vitro mutagenesis and identification of new variants via RAPD markers for improving *Chrysanthemum morifolium*. African Journal of Agricultural Research. 2010;**5**:748–757. DOI: 10.5897/AJAR09.679

[38] Xi M, Sun L, Qui A, Liu J, Xu J, Shi J. In vitro mutagenesis and identification of mutants via ISSR in lily (*Lilium longiflorum*). Plant Cell Reports. 2012;**31**:1043–1051. DOI: 10.1007/s00299-011-1222-8

[39] Bala M, Singh KP. In vitro mutagenesis of rose (*Rosa hybrida* L.) explants using gamma-radiation to induce novel flower colour mutations. The Journal of Horticultural Science and Biotechnology. 2015;**88**:462–468. DOI: 10.1080/14620316.2013.11512992

[40] Perera D, Barnes DJ, Baldwin BS, Reichart NA. Mutagenesis of in vitro cultures of *Miscanthus* × *giganteus* cultivar Freedom and detecting polymorphisms of regenerated plants using ISSR markers. Industrial Crops and Products. 2015;**65**:110–116. DOI: 10.1016/j.indcrop.2014.12.005

[41] Rai MK, Kalia RK, Singh R, Gangola MP, Dhawan AK. Developing stress tolerant plant through in vitro selection – An overview of the recent progress. Environmental and Experimental Botany. 2011;**71**:89–98. DOI: 10.1016/j.envexpbot.2010.10.021

[42] Biswas MK, Xu Q, Deng XX. Utility of RAPD, ISSR, IRAP and REMAP markers for the genetic analysis of *Citrus* spp. Scientia Horticulturae. 2010;**124**:254–261. DOI: 10.1016/j.scienta.2009.12.013

[43] Dhakshanamoorthy D, Selvaraj R, Chidambaran ALA. Induced mutagenesis in *Jatropha curcas* L. using gamma rays and detection of DNA polymorphism through RAPD marker. Comptes Rendus Biologies. 2011;**34**:24–30. DOI: 10.1016/j.crvi.2010.11.004

[44] Sen A, Alikamanoğlu S. Analysis of drought-tolerant sugar beet (*Beta vulgaris* L.) mutants induced with gamma radiation using SDS-PAGE and ISSR markers. Mutation Research. 2012;**738–739**:38–44. DOI: 10.1016/j.mrfmmm.2012.08.003

[45] Correia S, Matos M, Ferreira V, Martins N, Gonçalves S, Romano A, Pinto-Carnide O. Molecular instability induced by aluminum stress in Plantago species. Mutation Research. 2014;**770**:105–111. DOI: 10.1016/j.mrgentox.2014.06.002

[46] Christopoulos MV, Rouskas D, Tsantili E, Bebeli PJ. Germplasm and diversity and genetic relationships among walnut (*Juglans regia* L.) cultivars and Greek local selections revealed by Inter-Simple-Sequence Repeat (ISSR) markers. Scientia Horticulturae. 2010;**125**:584–592. DOI: 10.1016/j.scienta.2010.05.006

[47] Baliyan D, Sirohi A, Kumar M, Kumar V, Malik S, Sharma S, Sharma S. Comparative genetic diversity analysis in chrysanthemum: A pilot study based on morpho-agronomic traits and ISSR markers. Scientia Horticulturae. 2014;**167**:164–168. DOI: 10.1016/j.scienta.2013.12.029

[48] Roy A, Bandyopadhyay A, Mahapatra AK, Ghosh SK, Singh NK, Bansal KC, Koundal KR, Mohapatra T. Evaluation of genetic diversity in jute (Corchorus species) using STMS, ISSR and RAPD markers. Plant Breeding. 2006;**125**:292–297. DOI: 10.1111/j.1439-0523.2006.01208

[49] Sianipar NF, Ariandana, Maarisit W. Detection of gamma-irradiated mutant of rodent tuber (*Typhonium flagelliforme* Lodd.) in vitro culture by RAPD molecular marker. Procedia Chemistry. 2015;**14**:285–294. DOI: 10.1016/j.proche.2015.03.040

[50] Wu L, Li M, Yang X, Yang T, Wang J. ISSR analysis of chlorophytum treated by three kinds of chemical mutagen. Journal of Northeast Agricultural University. 2011;**18**(4):21–25. DOI: 10.1016/S1006-8104(12)60020-8

[51] Çelik O, Candar-Çakır B, Erdogmuş M. Comparative genetic analyses of salt tolerant soybean mutants by RAPD and ISSR. Unpublished data. Forthcoming.

[52] Bonifacio C, Santos IC, Belo C, Cravador C. Single-strand conformation polymorphism (SSCP) analysis of alfaS1-casein, beta-casein and K-casein genes in Charnequeira Portuguese indigenous goat breed. Revista Archivas de Zootecnia. 2001;**50**:105–111.

[53] Paux E, Sourdille P, Mackay I, Feuillet C. Sequence-based marker development in wheat: Advances and applications to breeding. Biotechnology Advances. 2012;**30**:1071–1088. DOI: 10.1016/j.biotechadv.2011.09.015

[54] Manzanares C, Yates S, Ruckle M, Nay M, Studer B. Tilling in forage grasses for gene discovery and breeding improvement. New Biotechnology. 2016;**33**:594–603. DOI: 10.1016/j.nbt.2016.02.009

[55] Varshney RK, Nayak SN, May GD, Jackson SA. Next-generation sequencing technologies and their implications for crop genetics and breeding. Trends in Biotechnology. 2009;**27**:522–530. DOI: 10.1016/j.tibtech.2009.05.006

[56] Yang C, Wei H. Designing microarray and RNA-seq experiments for greater systems biology discovery in modern plant genomics. Molecular Plant. 2015;**8**:196–206. DOI: 10.1093/mp/ssu136

Chapter 7

Neutron-Stimulated Gamma Ray Analysis of Soil

Aleksandr Kavetskiy, Galina Yakubova,
Stephen A. Prior and Henry Allen Torbert

Additional information is available at the end of the chapter

http://dx.doi.org/10.5772/68014

Abstract

This chapter describes technical aspects of neutron-stimulated gamma ray analysis of soil carbon. The introduction covers general principles, different modifications of neutron-gamma analysis, measurement system configuration, and advantages of this method for soil carbon analysis. Problems with neutron-gamma technology in soil carbon analysis and methods of investigations including Monte-Carlo simulation of neutron interaction with soil elements are discussed further. Based on the investigation results, a method of extracting the "soil carbon net peak" from the raw acquired data was developed. The direct proportional dependency between the carbon net peak area and average carbon weight percent in the upper 10 cm soil layer for any carbon depth profile was shown. Calibration of the measurement system using sand-carbon pits and field measurements of soil carbon are described. Measurement results compared to chemical analysis (dry combustion) data demonstrated good agreement between the two methods. Thus, neutron-stimulated gamma ray analysis can be used for *in situ* determination of near-surface soil carbon content and is applicable for precision geospatial mapping of soil carbon.

Keywords: soil carbon analysis, neutron-stimulated gamma ray analysis, Monte-Carlo simulation, Geant4, soil carbon mapping

1. Introduction

1.1. System evolution and application

Neutron-gamma analysis is based on detection of gamma lines that appear due to neutron-nuclei interactions. Many nuclei can be detected and quantified by the presence of these characteristic gamma lines. State-of-the-art nuclear physics methodologies and instrumentation, combined with commercial availability of portable pulse neutron generators, high-efficiency gamma detectors, reliable electronics, and measurement and data processing software, have currently made the application of neutron-gamma analysis possible for routine measurements in various fields of study. For these reasons, material analysis using characteristic gamma rays induced by neutrons is more wide-spread today; e.g., threat material detection (explosives, drugs, and dangerous chemicals [1]), diamond detection [2], planetary science applications for obtaining bulk elemental composition information, soil elemental (isotopic) content and density distribution [3], archaeological site surveying and provenance studies [4, 5], elemental composition of human [6, 7] and animal [8, 9] bodies, real-time elemental analysis of bulk coal on conveyor belts [10, 11], chloride content of reinforced concrete [12, 13], and in oil well logging [14].

In addition to the aforementioned applications, neutron-stimulated gamma ray analysis can be used in soil science for *in situ* measurements of soil carbon. This method is based on detecting 4.44-MeV gamma rays issued from carbon nuclei excited by fast neutrons promptly after the interaction [15]. Accurate quantification of soil carbon is important since it is an indicator of soil quality [16] that can affect soil carbon sequestration, fertility, erosion, and greenhouse gas fluxes [17–20].

Use of this method for soil elemental analysis has additional advantages over traditional laboratory chemical methods. This is a nondestructive *in situ* method of analysis that requires no sample preparation and can perform multielemental analyses of large soil volumes that are negligibly impacted by local sharp changes in elemental content. These advantages support the use of the neutron-gamma method in soil science.

1.2. General principles of neutron-gamma analysis

Neutron-gamma analysis is based on nuclei issuing gamma rays upon interaction with neutrons (**Figure 1**). Gamma rays are issued due to different processes of neutron-nuclei interactions. First of all, there are inelastic neutron scattering (INS) and thermal neutron capture (TNC) where gamma rays are issued promptly after interaction. New radioactive isotopes can appear due to INS and TNC processes, and decay of these isotopes is accompanied by delay activation (DA) gamma rays.

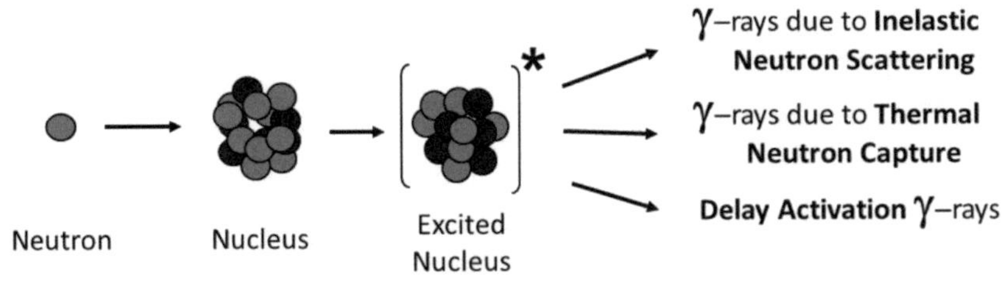

Figure 1. Main processes of neutron interaction with nuclei.

Each kind of nucleus and process produces gamma-rays of particular energy. In some cases, this characteristic gamma line of particular energy can serve as an analytical line for elemental determination. For some elements, the energy of characteristic gamma lines of nuclei and the processes responsible for the appearance of these gamma lines are listed in **Table 1**. As can be seen, gamma ray energy lies in the 1–11 MeV range. This is the range (greater than 1.022 MeV) where the effect of pair production as gamma rays interact with matter is significant. This is why

Element/nucleus	Applied for analysis			
	Kind of neutrons	Process	Cross section, b (neutron energy)	Characteristic gamma line, MeV
Silicon/^{28}Si	Fast	INS	0.52 (14 MeV)	1.78
Oxygen/^{16}O	Fast	INS	0.31 (14 MeV)	6.13
Hydrogen/^{1}H	Thermal	TNC	0.33 (thermal)	2.22
Carbon/^{12}C	Fast	INS	0.42 (14 MeV)	4.44
Nitrogen/^{14}N	Thermal	TNC	0.08 (thermal)	10.82
	Fast	INS	0.39 (14 MeV)	2.31, 4.46, 5.03, 5.10, 7.03

Table 1. Gamma lines used in neutron-gamma analysis of some elements.

the single escape (SE) and double escape (DE) peaks appear in the gamma spectra near the full energy peak [21]. For example, with carbon registration at full energy peak of 4.44 MeV, peak shifts of 0.511 MeV (SE, 3.93 MeV) and 1.022 MeV (DE, 3.42 MeV) can be observed. The cross sections of INS process for 14-MeV neutron interactions with nuclei are demonstrated in **Table 1**. For instance, the value of the ^{12}C cross section at neutron energy of 14.1 MeV is ~0.42 barn. The inelastic scattering of fast neutrons on ^{12}C nuclei elevates them to the 4.44-MeV exited energy state [22]. Exited state $^{12}C^{*}$ promptly returns to the ground state issuing the 4.44-MeV gamma ray.

$$n + {}^{12}C \rightarrow {}^{12}C^{*} + n' \rightarrow {}^{12}C + \gamma(4.44\ \text{MeV}) \tag{1}$$

Neutrons lose their energy when propagating through the medium. The interaction cross section depends on energy. The dependence of the INS process cross section with energy for ^{12}C is demonstrated in **Figure 2**. The intensity or peak area of this gamma line in the spectrum can be associated with the amount of carbon in a given soil volume. Thus, the registration of the gamma spectra from the studied object caused by neutron interaction with its nuclei can be used for elemental analysis of the object.

1.3. Measurement system configuration

The configuration of a measurement system for neutron-gamma analysis should consist of a neutron source, gamma detector, shielding and construction materials, operational electronics, and data acquisition software. Below we briefly consider the main features of these component parts.

1.3.1. Neutron sources

Isotope neutron sources (based on Cf-252, Am-241-Be, Pu-238-Be isotopes) and portable neutron generators can be used in the measurement setup; some commercially available neutron sources are listed in **Table 2**. Although radioisotope sources are widely used in neutron-gamma analysis [23–26], the use of a neutron generator is preferred (from a radiation safety point of view) since no radiation is produced when the generator is turned "off." Furthermore, the availability of pulse neutron generators has significantly expanded the possibilities of this method [1, 2, 11, 27].

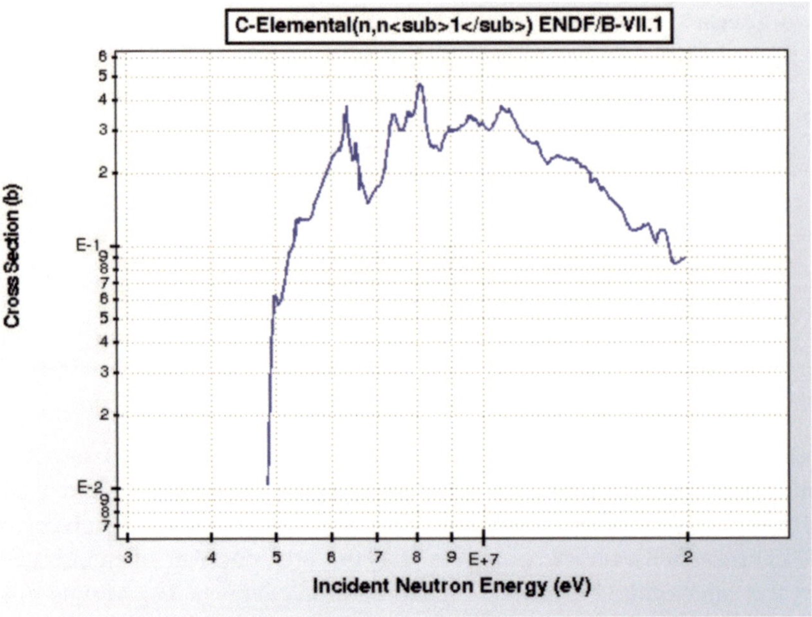

Figure 2. Inelastic neutron scattering cross section of ^{12}C nuclei [22].

Type of source		Nuclear reaction	Time of work = $T_{1/2}$ or working mode	Neutron energy, MeV, Avg (max)	Flux, n/s	Reference
Isotope	$^{241}Am/Be$	$^{241}Am \rightarrow \alpha + ^{237}Np$ $\alpha + ^{9}Be \rightarrow ^{12}C + n$	432.6 yr	4 (11)	4e7	[22, 28, 29]
	$^{239}Pu/Be$	$^{239}Pu \rightarrow \alpha + ^{235}U$ $\alpha + ^{9}Be \rightarrow ^{12}C + n$	24100 yr	4.5 (10.7)	4e6	[22, 28, 30]
	$^{210}Po/Be$	$^{210}Po \rightarrow \alpha + ^{206}Pb$ $\alpha + ^{9}Be \rightarrow ^{12}C + n$	138 d	4.2 (10.9)	2.5e6	[22, 28]
	^{252}Cf	spontaneous fission ^{252}Cf	2.65 yr (alpha decay)	2.3 (6)	4.4e7	[22, 28, 31]
	$^{226}Ra/Be$	$^{226}Ra \rightarrow \alpha + ^{222}Rn$ $\alpha + ^{9}Be \rightarrow ^{12}C + n$	1600 yr	3.9 (13.1)	1.5e7	[22, 28]
Neutron Generator	Genie 16	$d + d \rightarrow n + ^{3}He$	On-Off	2.5	2e8	[32]
	Genie 35	$d + d \rightarrow n + ^{3}He$	On-Off	2.5	1e8	[32]
		$d + t \rightarrow n + \alpha$		14.1	1e10	
	P 211	$d + d \rightarrow n + ^{3}He$	On-Off	2.5	1e8	[33]
		$d + t \rightarrow n + \alpha$		14.1		
	P 385	$d + d \rightarrow n + ^{3}He$	On-Off	2.5	5e8	[33]
		$d + t \rightarrow n + \alpha$		14.1		
	D 711	$d + d \rightarrow n + ^{3}He$	On-Off	2.5	2e10	[33]
		$d + t \rightarrow n + \alpha$		14.1		

http://dx.doi.org/10.5772/68014

Type of source	Nuclear reaction	Time of work = $T_{1/2}$ or working mode	Neutron energy, MeV, Avg (max)	Flux, n/s	Reference
MP320	d+d→n+^{3}He	On-Off	2.5	1e8	[33]
	d+t→ n+α		14.1		
	API 120	d+t→ n+ α	On-Off	14.1	2e7
[33]					
ING	d+d→n+^{3}He	On-Off	2.5	1e8	[34]
	d+t→ n+ α		14.1	1e10	

Table 2. Some neutron sources available for use in neutron-gamma analysis.

1.3.2. Gamma detectors

Gamma detectors used in neutron-gamma analysis systems should be suitable for operation in mixed radiation fields where neutrons and gamma rays are present. In ideal cases, detectors should have the following properties [1]:

- detector material must have a high Z value, and total detector volume should be quite large (more than 1 dm^3, preferably 5–10 dm^3) to effectively detect a relatively low flux of characteristic gamma rays with energy up to ~12 MeV;
- detectors must provide energy resolution that allows for resolving peaks of interest;
- the interaction of neutrons with nuclei in the detector material should not produce gamma rays that overlap useful signals emitted from samples; the detector material should be void of isotopes that are anticipated in analyzed samples;
- neutron activation of isotopes inside the detector volume with time-delayed radiation should be minimal;
- detector sensitivity to changing environmental conditions (temperature, etc.) should be minimal for field application of neutron-gamma analysis.

Satisfaction of these requirements can be difficult, especially due to budget constraints. Among the different types of gamma scintillators, inorganic scintillators are more suitable for neutron-gamma analysis systems due to higher efficiency of registering gamma rays in the energy range up to 12 MeV. The high sensitivity of inorganic scintillation detectors is assured by high gamma ray energy deposition in the relatively large volume (up to several cubic decimeters) of transparent inorganic gamma scintillator mono-crystals with a high Z and density and by their high light yield values (photons per MeV). Semiconductor detectors have a better resolution compared to scintillation detectors, but lower registration efficiency in the desired energy range (up to 12 MeV), which makes scintillation detectors more preferable for use in neutron-gamma analysis systems.

Properties of detectors [based on the sodium iodide NaI(Tl), bismuth germinate BGO, and lanthanum bromide $LaBr_3(Ce)$] commonly used in the neutron-gamma analysis systems are described in **Table 3**. Note that other detectors based on inorganic scintillators have worse characteristics and are not usually applied in neutron-gamma analysis.

Detector type	Light yield, photon/MeV	Scintillation decay time, ns	Resolution, % (at 662 keV)	Density, g/cm^3	Effective atomic number*	Reference
NaI(Tl)	38000	250	7	3.67	47	[35]
$Bi_4Ge_3O_{12}$ (BGO)	8200	300	10	7.13	62	[36, 37]
$LaBr_3$(Ce)	70000	17	3	5.07	43	[38, 39]

*Effective atomic number is calculated by [40].

Table 3. Properties of gamma detectors used in neutron-gamma analysis.

1.3.2.1. *NaI(Tl)*

As shown in **Table 3**, all listed detectors have high light yield. These detectors with sizes around dia 15 × 15 cm provide practically 90% adsorption of gamma rays with energy up to 10 MeV as shown by data in **Figure 3** for sodium iodide detectors. For other detectors, sizes can be less due to higher density and effective atomic numbers. Sodium iodide detectors under neutron irradiation are activated, showing the delayed beta decay spectral continuum with end point energy of 2 MeV [1]. But this activation by neutron fluxes in neutron-gamma analysis does not significantly impact the neutron-stimulated gamma spectra [41]. There are no significant differences in energy resolution before and after irradiation by 4.7×10^{11} of 14-MeV neutrons for the dia 10 × 10 cm NaI(Tl) detector [42]. The radiation damage of sodium iodide occurs at an adsorption dose of 500–1000 Gy [43], which is not accumulated in real time when conducting neutron-gamma analysis.

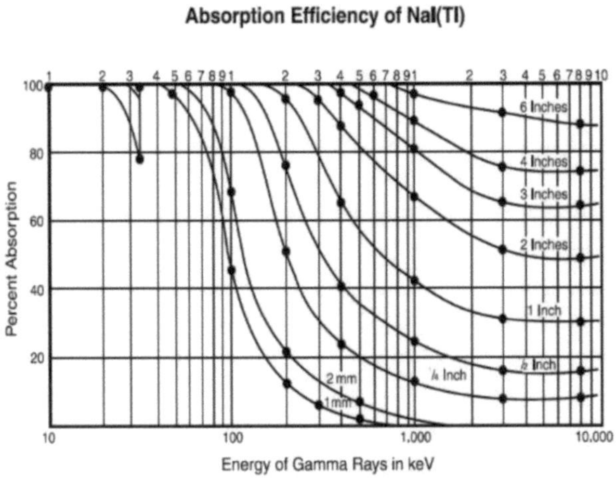

Figure 3. The family of curves derived from NBS circular 583 (1956), Table 37, mass attenuation coefficients for NaI(Tl). Each curve represents the percent absorption (I-attenuation) of a parallel beam of gamma rays normally incident on NaI (Tl) crystals of a given thickness [44].

1.3.2.2. BGO

The relatively small light yield of BGO scintillators is compensated by the higher densities and atomic numbers of the composition elements. BGO scintillators have approximately the same efficiency as NaI(Tl), but the interactions of neutrons with BGO elements result in the appearance of gamma lines with energy up to 2.5 MeV, which makes this detector unsuitable for measuring gamma spectra in this range in neutron-gamma analysis [45]. Also, a significant drawback of this type of detector is sensitivity to external temperature [46], but a thermal correction system can compensate for this disadvantage [47].

1.3.2.3. $LaBr_3(Ce)$

Among inorganic scintillators, $LaBr_3(Ce)$ demonstrated the best resolution and efficiency. Due to the shortest scintillation decay time, this detector had lower background in the high energy part of the spectra due to the smaller number of pile-ups of low energy photons. The presence in this detector of small quantities of the ^{138}La radioactive isotope produces a 1.47-MeV gamma peak, which is always visible in the gamma spectrum and can be used for calibration purposes [1], but does not significantly impact the neutron-stimulated spectra. This detector has stable gamma ray spectra parameters when properly shielded against direct neutron flux from the neutron source. It is the best candidate for active neutron applications, but the high cost of this detector (7.62 cm × 7.62 cm $LaBr_3(Ce)$ costs ~US$35,000 vs US$2,000 for a high quality NaI(Tl) of similar size [39]) limits a wider use compared to NaI(Tl) and BGO detectors.

1.3.3. Shielding and construction materials

Direct fast neutron flux on the gamma detector and gamma radiation appearing from neutron interaction with detector nuclei leads to high gamma spectra background. High background increases the minimal detection limit of the measurement system and measurement errors. Shielding use between the neutron generator and gamma detector improves characteristics of the measurement system.

Shielding that is one-layer [48–51] or multilayer [52–54] can be used for this purpose. In most cases, fast neutron shielding consists of two or three components which first slows fast neutrons to thermal energy (moderator), absorbs thermal neutrons (absorber), and then attenuates the gamma rays which are produced by different neutrons-nuclei interactions in the moderator and absorber (attenuator). Light materials like water, heavy water, and polyethylene are usually used as neutron moderators, and boric acid is a possible absorber. Sometimes iron is used in the first layer ahead of the light materials to moderate fast neutrons via inelastic neutron scattering [52, 53]. Borated water or borated polyethylene can serve as a combined moderator and absorber in the first layer of shielding. Lead, tungsten, iron, or other such materials with high atomic mass are used to decrease gamma radiation.

While decreasing background, the shielding material and geometry should allow for the counting of useful signal. This means that the shielding thickness should be reasonable and should not produce gamma lines within the energy range of interest. Additionally, it is important to know the possible high energy gamma lines produced from a particular shielding material since they could interfere with useful gamma lines or give additional continuous background lower energy due to the Compton Effect. Construction materials should have minimal susceptibility to neutron activation by fast or thermal neutrons, issue few gamma rays in the energy range of interest, and have minimum high energy gamma rays that increase system background.

1.3.4. Operational electronics and data acquisition software

Operational electronics and data acquisition software essentially depend on the task and particular method modifications. For example, prompt gamma neutron activation analysis with a radioactive isotope neutron source, a standard gamma detector, and multichannel analyzer (e.g., MCA-1000) with its own software could be used [23]. A complicated custom-made experimental setup consisting of standard Ortec or Canberra electronic blocks paired with a pulsed neutron generator and gamma and alpha detectors can be used for dangerous material detection (as described in [55]). A custom-made electronic scheme and data acquisition software could be used in some cases due to the absence of suitable standard equipment (e.g., NaI(Tl) detector with corresponding electronics and ProSpect v0.1.11-vega software from XIA LLC, Hayward, CA; see Ref. [56]).

1.4. Modifications of neutron-gamma analysis

Depending on the area of application, different modifications of neutron-gamma analysis can be used. Detailed descriptions of these methods were presented in Ref. [27]; we briefly list these methods below:

- PGNAA—Prompt Gamma Neutron Activation Analysis
- PFNA—Pulsed Fast Neutron Analysis
- PFTNA—Pulsed Fast/Thermal Neutron Analysis
- PFNTS—Pulsed Fast Neutron Transmission Spectroscopy
- API—Associated Particle Imaging

Pulsed Fast/Thermal Neutron Analysis (PFTNA) is the most suitable for soil neutron-gamma analysis [57]. The main difficulty conducting soil neutron-gamma analysis is the overlapping of different gamma lines from soil and measurement system nuclei and processes with the main peak of interest (e.g., soil carbon peak). The PFTNA system makes it possible to separate the gamma ray spectrum due to INS reactions (n,n′γ) from the TNC (n,γ) and DA reaction (e.g., (n,p)) spectra. The moderation and moving neutrons in matter limit the incoming neutron—matter nucleus reaction speed. Approximately 1.5 microseconds are required to moderate 14-MeV neutrons to thermal energy in hydrogenous materials [58], while the lifetime of thermal neutrons can be hundreds of microseconds [59]. Thus, INS reactions will only occur during the microsecond neutron pulse, while TNC processes are running during the neutron pulse and between pulses. One memory address of the data acquisition system records during the neutron pulse, while another memory address acquires data between pulses. This is a technique used with small portable electronic neutron generators (see Table 2). PFTNA employs pulses with a duration of 5–20 microseconds. Microsecond pulse durations significantly reduce PFTNA system cost compared to pulsed methods using nanosecond neutron pulses. The PFTNA system employs pulse frequencies greater than 5 kHz to ensure nearly constant thermal neutron flux for the measurement period [60]. When operating at 10 kHz and a 25% duty cycle neutron pulse, it was demonstrated that net count rates in the individual peaks of the soil elements silicon and oxygen in the TNC spectrum have a steady state between neutron pulses [61]. Thus, at first approximation, the count rate registration of gamma flux, which appears under neutron irradiation of samples, can be accepted.

2. Neutron-gamma technology for soil carbon determination

2.1. Importance of soil carbon determination

Adoption of agricultural land use practices adapted for climate change and mitigation potential depends on agricultural productivity and profitability. Understanding and evaluating the impacts on soil resources will influence the development of sustainable land use practices. A critical component of any soil resource evaluation process is measuring and mapping natural and anthropogenic variations in soil carbon storage. Soil carbon can impact many environmental processes, such as soil carbon sequestration, fertility, erosion, and greenhouse gas fluxes [17–20]. The current "gold standard" of soil carbon determination is based on the dry combustion technique (DCT) [62]. This method is destructive, time-consuming, and labor-intensive since it involves collecting extensive field soil core samples and requires lots of sample preparation before complex laboratory analysis can be conducted. Furthermore, DCT soil analysis represents a point measurement in space and time that cannot be confidently extrapolated to field or landscape scales which limits its utility for expansive coverage or longer timescale interpretation. Other techniques include laser-induced breakdown spectroscopy, near- and mid-infrared spectroscopy, diffuse reflectance infrared Fourier transform spectroscopy, and pyrolysis molecular beam mass spectrometry [15].

Soil neutron-activation analysis is a new method with the potential for measuring soil carbon in relatively large volumes without having to take destructive soil samples requiring time-consuming standard laboratory analysis. This new method is based on measuring the gamma response of soil irradiated with fast neutrons. One modification of this method, PFTNA—Pulsed Fast/Thermal Neutron Analysis, has been shown to provide wide-area monitoring for prolonged periods [15, 53]. The result of measurements using this method gives, as will be demonstrated below, the values of average carbon content in weight percent in the upper soil layer (thickness ~10 cm) of ~1.5 m^2 area centered under the neutron source. The measurement time for each surveyed area is 30–60 minutes.

2.2. Problems with neutron-gamma technology in soil carbon analysis and methods of investigation

2.2.1. Features of soil carbon neutron-gamma analysis

The main purpose of this book chapter is to describe the application of neutron-gamma technology for soil elemental analysis. Common features of this technology were described earlier in the introduction. The following aspects of neutron-stimulated gamma ray analysis will be covered:

- mobile neutron-gamma technology systems for soil carbon content determination;
- procedures for measuring the gamma response of neutron-irradiated soil (raw data) and extracting the net soil carbon signal;
- soil carbon depth distribution and the particular soil carbon characteristics that are directly and proportionally dependent on the net carbon signal;
- comparison of neutron-gamma field measurements of soil carbon content to traditional chemical analysis;
- factors impacting gamma response intensity in quantitative soil analysis.

2.2.2. *Methods of investigation*

The methods used for investigating the effects of different factors when applying neutron-gamma technology for soil elemental analysis are:

- Experimental design. Soils being experimentally measured should be around a cubic meter in volume and weigh around a metric ton.
- Deterministic modeling. This method involves solution of integral or differential equations that describe the dependence of behavioral characteristics of the system in question in terms of spatial or time coordinates. This method was used in cases of simple shapes and sample properties (e.g., uniform distribution of elements within the sample volume). This method gives useful semi-quantitative results.
- Monte-Carlo (MC) simulation. The gamma response spectra from modeled soil samples irradiated by neutrons are a very effective method to determine the effect of different factors on the neutron-gamma measurement. An MC simulation model of any sample shape, shielding and detector configuration, or measurement geometry is applicable. MC simulation results are very close to experimental findings.

All these methods were used during our investigations. Results from these methods will be discussed and compared with each other.

2.2.3. *Monte-Carlo simulation method*

MC simulations [63, 64] have been extensively used to solve various problems. For example, MC simulations are capable of estimating the neutron flux passing through materials and their energy loss in these materials, determining the energy distribution of emerging neutrons [65], calculating the optimal thickness of shielding [66, 67] and moderator [68], and reproducing the characteristic neutron-induced gamma-ray spectra of different materials [69–73].

An MC simulation model of soil neutron-gamma analysis should consist of two major components—the measurement system and a soil model. The modeled measurement system should mimic the experimental measurement system. The system could have neutron sources (isotropic source with energies that match the experimental setup), detector, and shielding (if required). The MC simulation soil model can be viewed as a three-phase system (solid, liquid, and gaseous phases) [74]. Based on calculation objectives, the soil model may be simplified if all soil parameters critical to the MC simulation are met. Our research used the approach of other researchers [74, 75] where the soil model was constructed as a compact medium with known elemental composition and density depth profiles.

The MC simulation describes randomly issued neutron transportation which includes all interactions with soil components until reaching the simulation volume boundary or exhausting its kinetic energy and disappearing due to an interaction. Some neutron-nuclei interactions result in the appearance of gamma rays which move through and interact with soil components. These interactions cause gamma quanta to disappear as they lose energy; however, some will propagate through the soil and be counted by the detector. The simulated gamma spectrum represents the relationship of the gamma count versus energy. The spectrum shape (number of peaks, their intensity) will be influenced by soil properties. The variation of modeled soil properties and MC simulation of the gamma spectra makes it possible to detect the effect of different soil parameters on the shape of the spectra. Note that the MC simulation gave results that were very close to real data.

2.3. Mobile system for soil carbon determination

As previously described [56], our PFTNA system was mounted on a platform that could be transported by tractors or all-terrain vehicles over various field terrains. The dimensions of our mobile platform were 75 cm × 23 cm × 95 cm and weighs ~300 kg. While the primary construction material was aluminum, the iron shielding contributed more weight. Previous findings [53, 76] were used as a basis for the current construction and electronic system requirements. Our PFTNA system had three separate construction blocks (**Figure 4**). Components of the first block were an MP320 pulsed neutron generator (Thermo Fisher Scientific, Colorado Springs, CO), an R2D-410 neutron detector (Bridgeport Instruments, LLC, Austin, TX), and a power system (**Figure 4a, d**). The neutron generator has a pulsed output of 10^7–10^8 n s^{-1} (depending on parameter settings) and neutron energy of 14 MeV. Components of the PFTNA power system were four DC105-12 batteries (12 V, 105 Ah), a DC-AC inverter (CGL 600W-series; Nova Electric, Bergenfield, NJ), and a Quad Pro Charger model PS4 (PRO Charging Systems, LLC, LaVergne, TN). The first block also contained water, iron and boric acid shielding for isolating the detector from the neutron beam and focusing the beam on the soil area of interest. The second block had the gamma ray measuring equipment (**Figure 4b, e**) and contained three 12.7 cm × 12.7 cm× 15.2 cm scintillation NaI(Tl) detectors (Scionix USA, Orlando, FL) with corresponding XIA LLC electronics (XIA LLC, Hayward, CA). For equipment operation, the third block (**Figure 4c**) housed a laptop computer for controlling the neutron generator, detectors, and data acquisition system ProSpect 0.1 (XIA LLC) (**Figure 4c**).

In our applied PFTNA technique, gamma rays emitted by soil chemical elements under pulsed neutron irradiation were divided into two groups: emissions during the neutron pulse due to INS and thermo-neutron capture (TNC) and emissions between neutron pulses due to TNC reaction. Delay gamma rays (i.e., caused by neutron activation reactions) are also captured in these spectra. The two concurrent gamma spectra from each PFTNA measurement (i.e., INS +TNC and TNC spectra) were treated together. Spectra acquisition from the three gamma

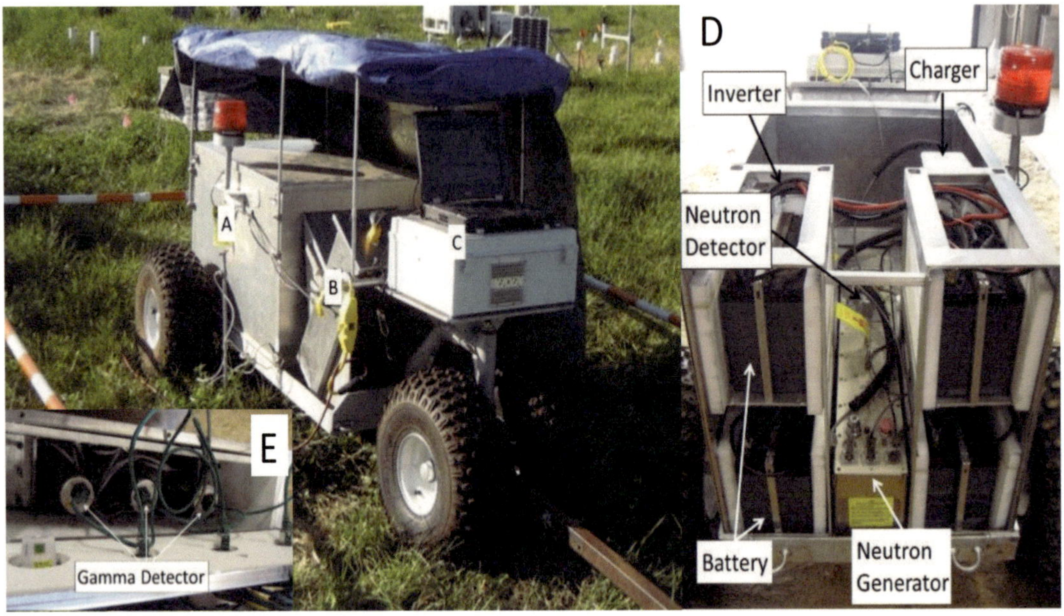

Figure 4. Overview of the PFTNA sytem: (a) neutron generator, neutron detector, and power system; (b) three NaI (Tl) detectors; (c) equipment operation; (d) general view of A showing individual components; and (e) close-up view of the gamma detectors [56].

detectors can be performed in two separate ways. In the first, analog signals from the detectors go to a summing amplifier for processing by a digital multichannel analyzer [77]. In the second, each detector has a dedicated analog-digital converter for spectra acquisition which can be summarized after correction for energy calibration instability [76, 78]. Our testing showed improved resolution from the second method which was therefore adopted for use in our PFTNA. For autonomous operation under field conditions, we developed a mobile power system for reliable equipment operation over extended periods of time. In this mobile PFTNA system, the neutron generator, neutron and gamma detectors, and laptop computer were all powered by four batteries via a power inverter. This inverter transformed 12 VDC battery power to 110 VAC and could operate with input voltages between 10.9 and 14.7 V.

2.4. Raw data acquisition

Two gamma spectra are acquired with our PFTNA measurements: (1) inelastic neutron scattering (INS) spectra acquired during the neutron pulse and (2) thermo-neutron capture (TNC) spectra acquired between neutron pulses. Typical INS and TNC experimental gamma spectra from soil (raw spectra) are shown in **Figure 5** (top and bottom lines, respectively). Each spectrum has a background spectrum and lines due to gamma emission from neutron-irradiated soil elements. The main gamma peak of interest has a centroid at 4.45 MeV in the INS spectrum. This peak may be due to neutron interactions with carbon nuclei and the interference of gamma lines from other nuclei. The oxygen peak (6.13 MeV) and the pair production peak (0.511 MeV) are used as reference points for spectral calibration. The INS spectra consist of gamma rays appearing from inelastic neutron scattering, thermal neutron capture, and delay activation of nuclei (samples and system construction materials). The TNC spectra consist of gamma rays from all of the above-listed processes except the INS process.

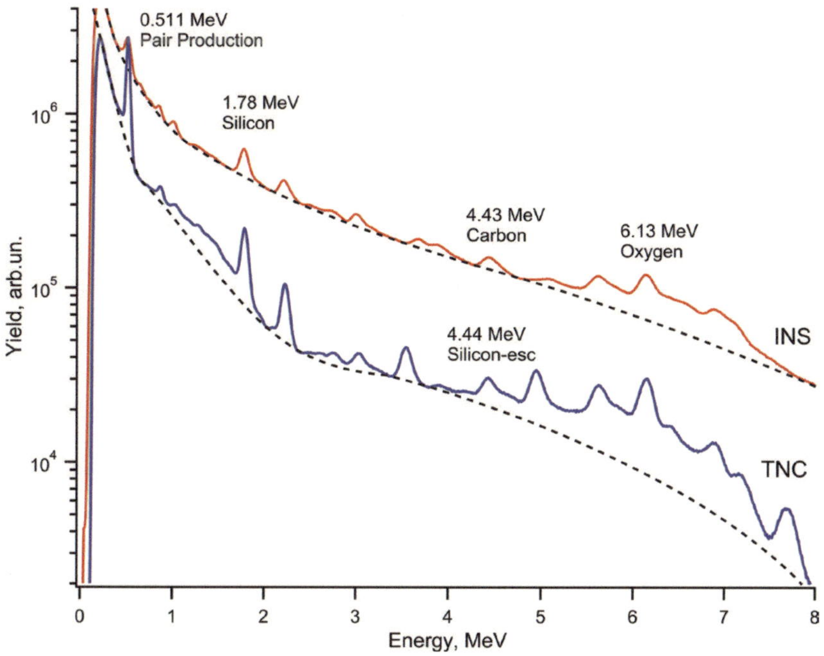

Figure 5. Raw experimental soil gamma spectra [56].

2.5. Extracting the "Net INS Spectra"

The first step in data processing is extracting the "net INS spectra" from raw data. The acquisition time of INS spectra is the duty cycle of neutron pulses multiplied by measurement clock time (minus dead time of multichannel analyzer), while the acquisition time of TNC spectra is one minus the duty cycle of neutron pulses multiplied by time of measurement (minus dead time of multichannel analyzer). Information on acquisition times is made available by the data acquisition software. Thus, spectra can be represented in counts per second (cps). As a first approximation, the number of INS events in some sample volume at some time moment is proportional to the number of fast neutrons in this volume, while the number of TNC events in some sample volume at some time moment is proportional to the number of thermal neutrons in this volume. In the first approximation, the time dependence of the number of fast neutrons $n_f(t)$ can be estimated according to the equation:

$$\frac{dn_f(t)}{dt} = N(t) - \frac{n_f(t)}{\tau_f} \tag{2}$$

where t is a time, $N(t)$ is the neutron flux to the sample from neutron sources, s^{-1}, τ_f is the fast neutron moderation time, s. The fast neutrons convert to thermal neutrons at moderation. The time dependence of the thermal neutron number $n_{th}(t)$ can be estimated as:

$$\frac{dn_{th}(t)}{dt} = \frac{n_f(t)}{\tau_f} - \frac{n_{th}(t)}{\tau_{th}} \tag{3}$$

where τ_{th} is the lifetime of thermal neutrons. As was discussed earlier (see Section 1.4), the fast 14-MeV neutron thermalization time can be accepted as equal to 1.5 microseconds, while the thermal neutron lifetime can be equal to ~1000 microseconds. The pulse neutron flux (in neutrons per microsecond) with time can be described by the equation:

$$N(t) = \begin{matrix} 40 \; if \; floor\left(\frac{t}{200}\right) \le \frac{t}{200} < floor\left(\frac{t}{200}\right) + 0.25 \\ 0 \; otherwise \end{matrix}, \tag{4}$$

if the neutron flux is 10^7 neutrons per second, frequency is 5000 Hz (pulse time is 200 microseconds), and duty cycle is 0.25 (these pulse neutron generator working regime parameters are for PFTNA of soil).

Solutions for these simple model equations are presented in **Figure 6**. As can be seen, in the frame of this model the time dependence of fast neutron numbers in the sample practically coincides with neutron flux time dependence (**Figure 6a,b**), while the time dependence of the thermal neutron is saw-shaped (**Figure 6c**). If the average value of this "saw" increases at the beginning, the average value reaches a constant value after more than 5000 microseconds (**Figure 6d**). When the "saw" reaches a constant value, the increase in thermal neutrons during the neutron pulse is practically linear with time, and the decrease in thermal neutrons between pulses is also linear (see **Figure 6c**). For this reason, the average value of TNC events and consequently the average TNC gamma flux during the neutron pulses is equal to the average value of TNC events and average TNC gamma flux between the neutron pulses. Hence it is possible to accept that the TNC spectra intensity between pulses is approximately the same as the TNC spectra intensity during pulses (in cps per channel). Based on this consideration, the "net INS spectra" can be restored with channel-by-channel subtraction of the TNC spectra from the INS spectra (both expressed in cps).

2.6. Measurement system background signal

Net INS spectrum represents the gamma rays appearing due to inelastic neutron scattering in both the sample and PFTNA system construction materials. The spectrum due to INS of the system construction materials is the background signal of the measurement system. To measure this background signal, the system has to be spatially removed from large objects (e.g., ground, floor, walls, building ceilings). To achieve this, the PFTNA system could be raised above the ground and away from buildings and large objects by using a crane (**Figure 7**). The measured INS and TNC spectra at different heights above the ground are shown in **Figure 8**; "net-INS spectra" (difference between INS and TNC spectra, both in cps) are shown in **Figure 9**. The peaks in these spectra can be attributed solely to INS processes. Intensities can be evaluated to determine the height at which the signal remained uniform with no change. This "no change" signal is considered to be the net INS system background spectrum.

http://dx.doi.org/10.5772/68014

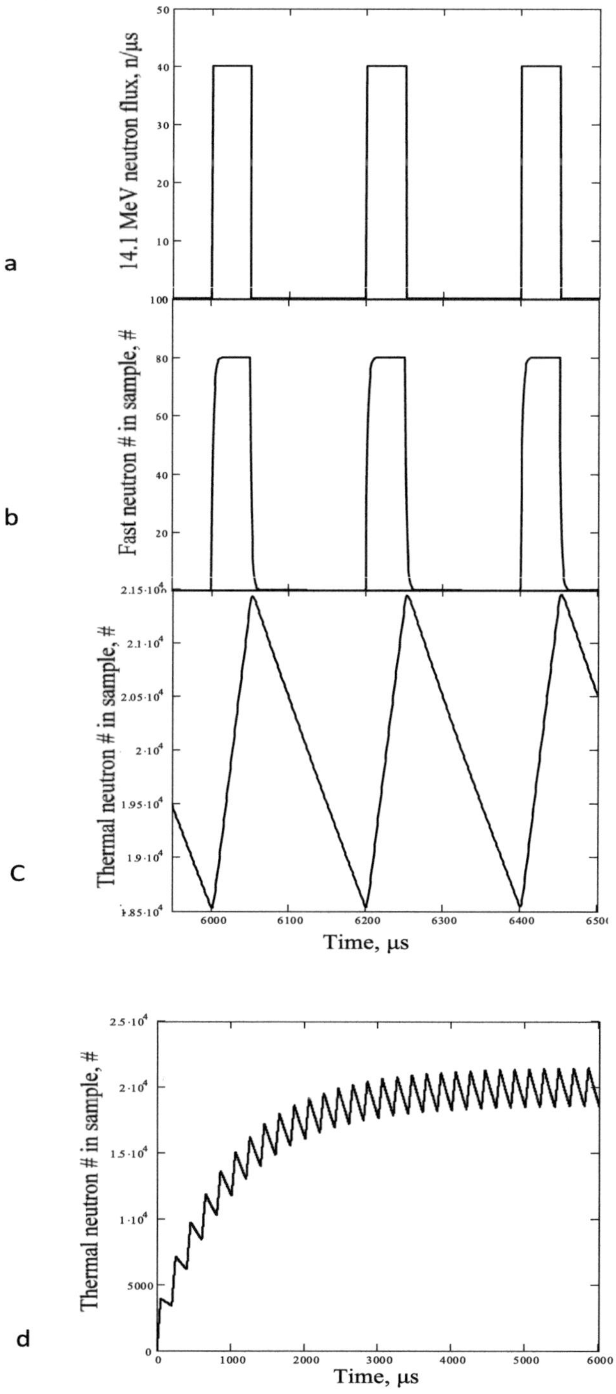

Figure 6. Time dependence of the neutron flux (a), number of fast neutrons in a sample $n_f(t)$ (b), number of thermal neutrons in a sample $n_{th}(t)$ (c), and time dependency of the number of thermal neutron in a sample at a time more less than 6000 microseconds (d).

Figure 7. The PFTNA system background measurements (up to 6.7 m above the ground) [79].

The behavior of peak areas with centroids at 1.78 MeV ("silicon peak," ^{28}Si), 4.45 MeV ("carbon peak"), and 6.13 MeV ("oxygen peak," ^{16}O) acquired from the net INS spectra are shown in **Figure 10**. As shown in **Figures 8–10**, some peaks in the spectra decrease and fully disappear with increasing height (e.g., peaks with centroids at 4.95 and 4.44 MeV in the TNC spectra), while other peaks decrease and reach constant values as height increases. Starting at ~4.5 m height, minimal spectral changes are detected. At this height, the measurement system

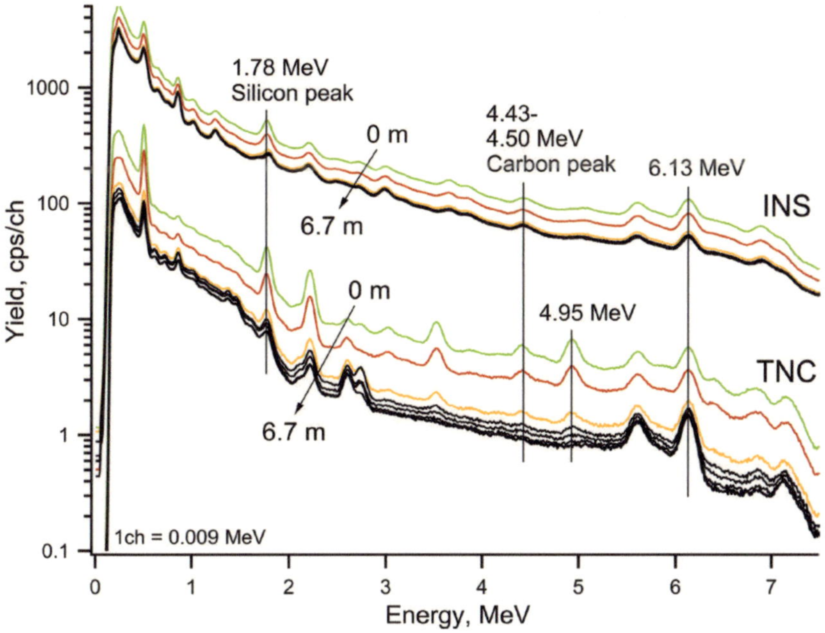

Figure 8. INS and TNC spectra measured by the PFTNA system at different heights above the ground [79].

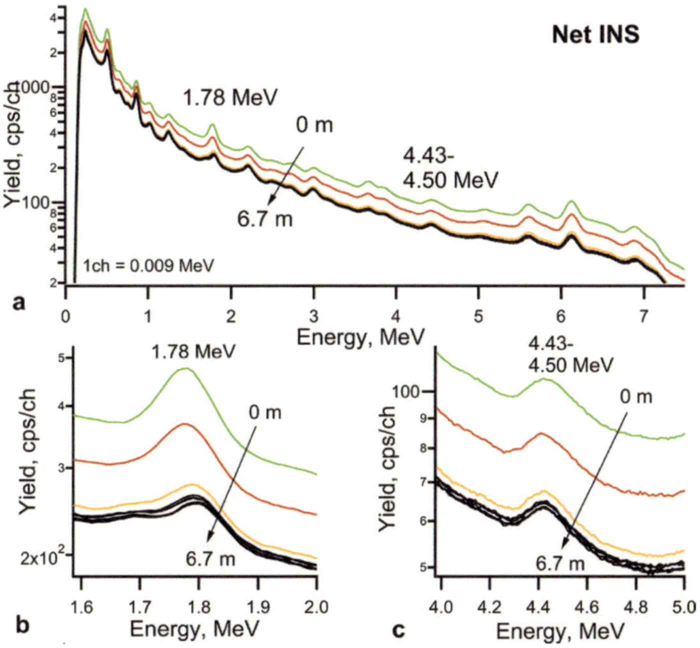

Figure 9. (a) Measurement system net INS spectra (difference between INS and TNC spectra, both in cps) at different heights above the ground; (b) fragment of the net INS spectra around 1.78 MeV; and (c) fragment of the net INS spectra around 4.43 MeV [79].

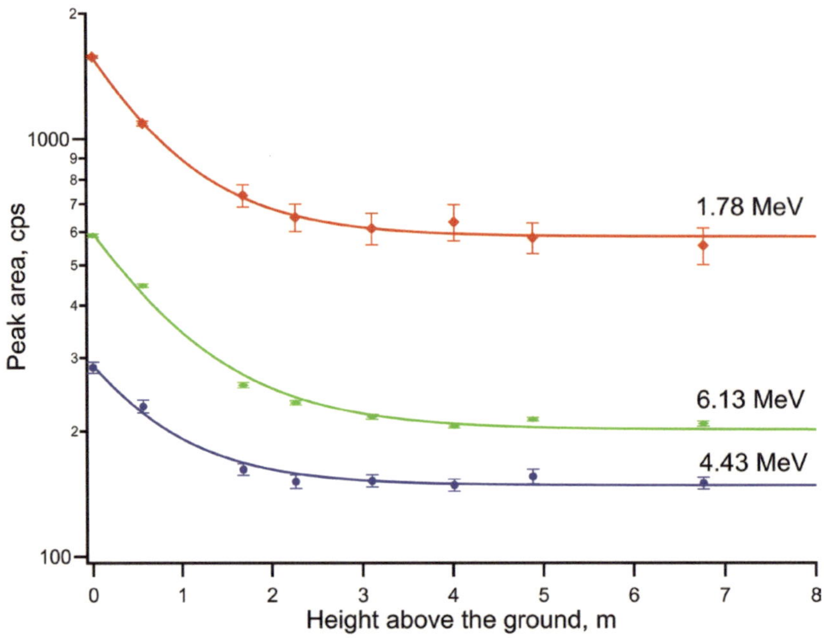

Figure 10. Dependencies of peaks areas with centroids at 1.78, 4.43, and 6.13 MeV in the net INS spectra for measurement system with changing heights above the ground [79].

is far enough away from the ground (and other large objects) that the gamma responses from these objects are negligible compared with the gamma responses from the measurement system construction materials. The net INS spectrum acquired at a height more than 4.5 m could be used as the system background spectrum.

2.7. "Soil Net INS Spectra" and "Soil Carbon Net Peak"

The "soil net INS spectrum" can be obtained from the results of soil measurements and the system background spectra. For this, "the system background net INS spectrum" should be subtracted (channel by channel) from the soil net INS spectrum received from the raw INS and TNC spectra (all spectra should be in cps). The "soil net INS spectrum" consists of gamma rays which appear due to inelastic neutron scattering of fast neutrons on soil nuclei.

Main peak of interest in "the soil net INC-spectra" is the peak with a centroid at 4.45 MeV. Analysis showed (see **Figure 11**) that this peak can consist of the soil carbon peak with centroid at 4.44 MeV, soil silicon cascade transition peak with centroid at 4.50 MeV; possibly the carbon peak with centroid at 4.44 MeV has contribution from excited carbon nuclei as a result of INS on other soil nuclei (e.g., due to ^{16}O (n,n′α)$^{12}C^{*} \rightarrow ^{12}C + \gamma$(4.44 MeV) reaction [80]). Silicon ^{28}Si nuclei turn to different excited states due to INS on silicon nuclei. The relaxation of excited silicon passes through the first exited state with energy 1.78 MeV, and the transition to ground state is accompanied by issued gamma rays with energy close to 1.78 MeV (i.e., "soil net silicon peak"). The relaxation of the 6.28 MeV silicon excited state pass to ground state through the first excited state and is accompanied by gamma rays with energy close to 4.50 MeV (6.28 − 1.78 MeV); that is the "silicon cascade transition peak" [22]. This peak can be a part of the 4.45 MeV

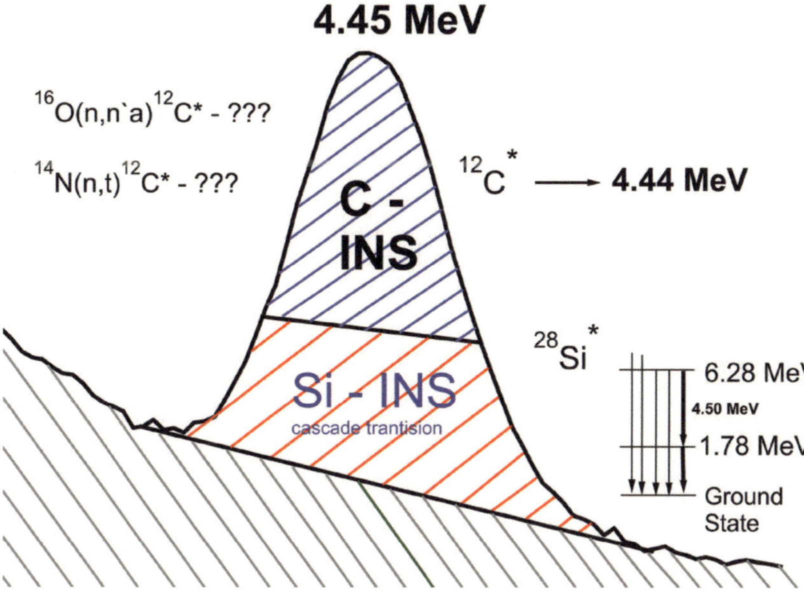

Figure 11. Composition of the 4.45-MeV peak in the soil net INS spectrum.

peak in the "soil net INS spectra." The theoretical calculation of the 4.50 to 1.78 MeV gamma ray intensity ratio (i.e., "cascade transition coefficient") gives a value of 0.0547 [81].

2.8. Defining "Soil Carbon Net Peak Area" for a uniform carbon depth profile

2.8.1. Measured gamma spectra of sand-carbon pits

Measurements of INS and TNC spectra using the PFTNA system were performed over 1.5 m × 1.5 m × 0.6 m pits filled with uniform sand-carbon mixtures that had carbon contents of 0, 2.5, 5, and 10 w%. The measurement system was placed over each pit such that the neutron source was situated over the geometric center of each pit. The "soil net INS spectra" were calculated for each pit, taking into account the system background spectra as described above. The experimental "net INS spectra" for pits are shown in **Figures 12** and **13**.

2.8.2. Monte-Carlo simulated gamma spectra of sand-carbon pits

MC simulations of gamma spectra from pits (1.5 m × 1.5 m × 0.6 m) with different sand-carbon mixtures using model geometry very similar to experimental system geometry were evaluated. The soil models are represented as compact media with above-mentioned dimensions and uniform SiO_2+C composition densities

$$d_{mix} = \frac{1.7 \cdot 0.52 \cdot 100}{Cw\% \cdot 1.7 + (100 - Cw\%) \cdot 0.52} \tag{5}$$

where 1.7 is the sand bulk density (g cm^{-3}); 0.52 is the coconut shell bulk density used in the pits as carbon (g cm^{-3}), and $Cw\%$ is the carbon content of the mixture in weight percent. The

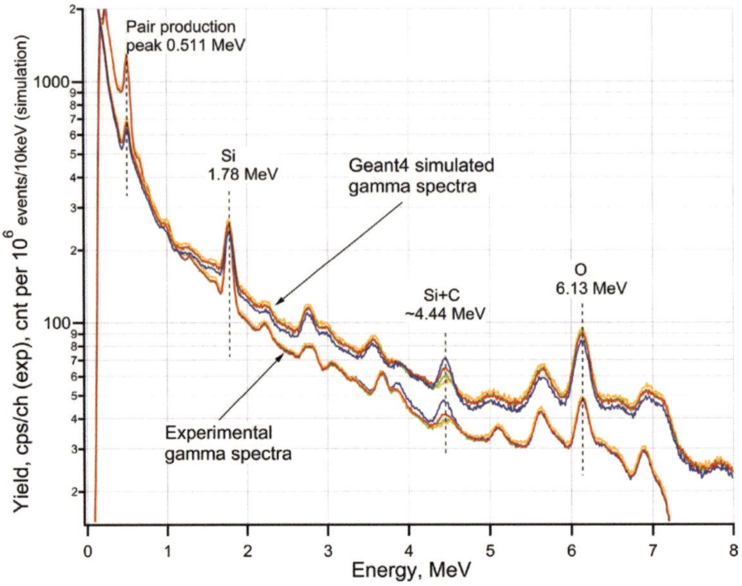

Figure 12. Simulated and measured 14-MeV neutron-stimulated net INS gamma spectra of sand-carbon mixtures (0, 2.5, 5, 10 w% C) in 1.5 m ×1.5 m × 0.6 m pits [79].

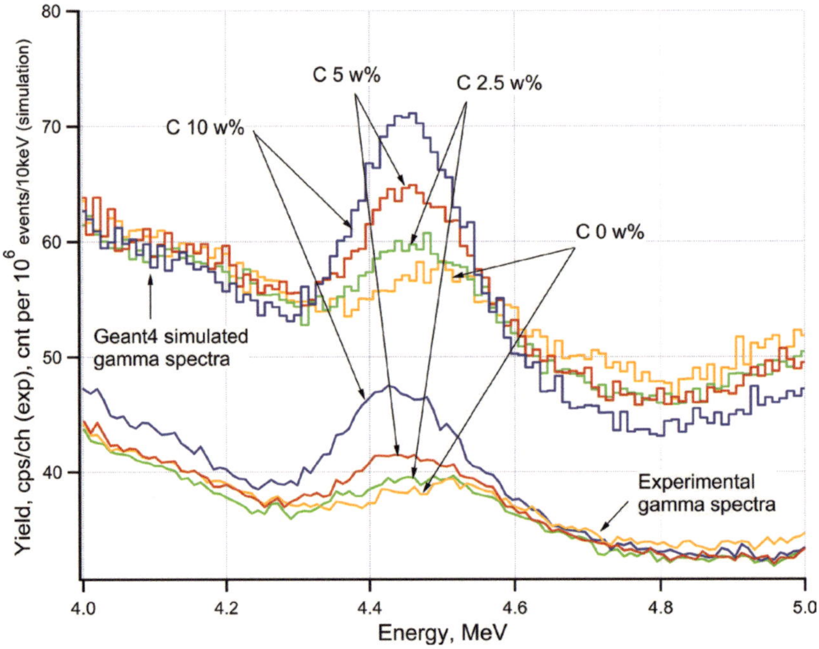

Figure 13. Fragment of simulated and measured 14-MeV neutron-stimulated net INS gamma spectra of sand-carbon mixtures (0, 2.5, 5, 10 w% C) in 1.5 m ×1.5 m × 0.6 m pits [79].

simulation model consisted of a point isotropic neutron source, gamma detector, and shielding similar to the real measurement system. The distance between the source and detector (35 cm), height of the model system above the ground, and number and type of detectors (three NaI(Tl) 12.7 cm × 12.7 cm × 15.3 cm) were the same as in the experimental system. The Geant4 tool kit [82] version G4.10.01p.01 [83] was used to conduct the MC simulations for this and other research issues. A conventional laptop with a multicore processor and high performance computing cluster (Auburn University Samuel Ginn College of Engineering vSMP HPCC consists of 512 cores @ 2.80 GHz X5560, 1.536TB shared memory, and 20.48TB raw internal storage) were used for calculation in the multithread mode. Note that for accuracy of the simulated spectra to approximately equal the experimental accuracy, 10^9 simulation events should be performed. Due to the large number of simulation events, the simulation time for each spectrum was several dozen hours. Our simulation used the neutron cross section JENDL4.0 database rather than the default database (G4NDL4.5) due to the JENDL4.0 simulated spectra and the experimental spectra being more similar. From Geant4 toolkit, we used the QGSP BIC HP and QGSP BERT HP physics lists (Reference Physics Lists, 2014). Both lists had high precision models for neutron transport below 20 MeV and gave the same simulation results. The change in detector energy resolution was taken into account as ~$1.142\cdot\sqrt{E_\gamma}$ (E_γ is the gamma quanta energy, keV) during simulation. This type of energy resolution dependence for gamma detectors is known [84], and the multiplier 1.142 was determined by matching the width of the simulated ^{137}Cs peak to that in the experimental spectra. The detector efficiency dependence with energy was not accounted for since this change would be minor due to the large NaI crystal sizes in the 1–10 MeV energy range [44]. Only INS spectra were simulated; other processes (like thermal neutron capture) in the simulation code were deactivated. A computer screenshot of the simulation model is shown in **Figure 14**.

2.8.3. MC simulated system background "INS spectra"

For determination of "pit net INS spectra," the system background "INS spectra" should first be simulated. In this simulation, the measurement model geometry and system components (detectors, shielding, sizes) were the same, but the pit model was absent. The "pit net INS spectra" are represented with channel-by-channel differences (simulation channel width is 10 keV) between "pit INS spectra" and system background "INS spectra."

The effect of system background on the simulated spectra is demonstrated as follows. **Figure 15** represents simulated "INS spectra" of the SiO_2+5w%C pit measured by a system consisting of a neutron source, sodium iodide detector, and different shielding. The system background spectra with different shielding are also shown in this plot. The "pit net INS spectra" (difference between "pit INS spectra" and "background INS spectra") are presented in **Figure 16**. As can be seen, the shape of each "pit INS spectra" measured by the system with different shielding is also different. Similar variations were also seen in "background INS spectra" for different shielding, but the "pit net INS spectra" were the same in all cases (**Figure 16**). Although these results may appear trivial, this example demonstrates that parts of system background in the raw spectra can be significant and should be taken into account by subtraction in quantitative analysis. Similar subtractions should be performed in experimental measurements.

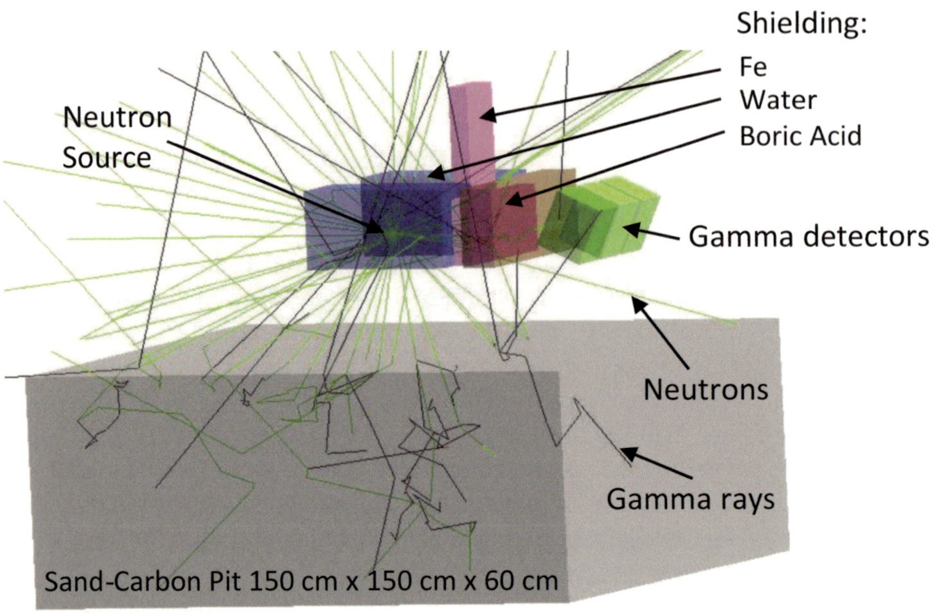

Figure 14. The MC simulation model (Geant4) for measurement of gamma response from a sand-carbon pit under neutron irradiation.

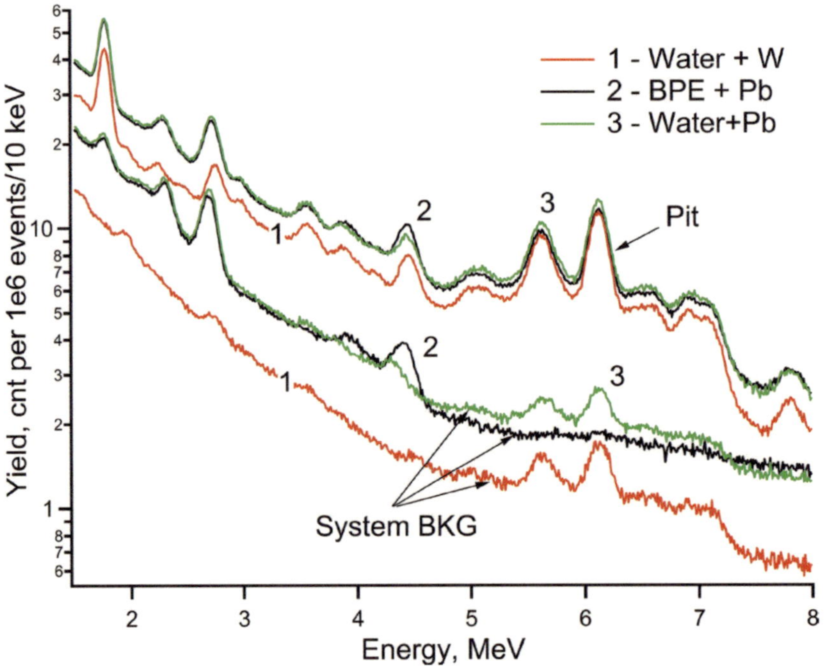

Figure 15. Simulated Geant4 gamma spectra of Pit SiO_2 + 5%C and background for system with different shielding.

http://dx.doi.org/10.5772/68014

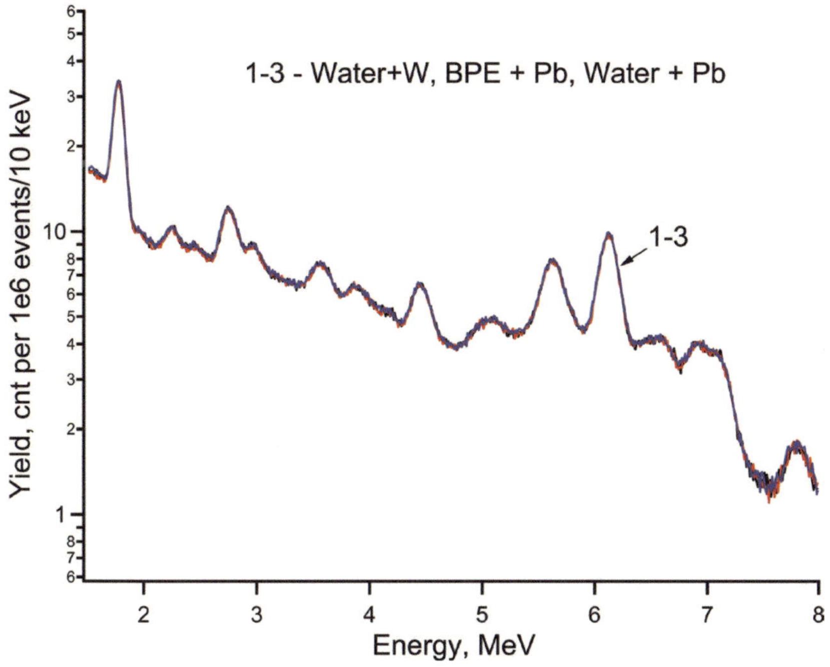

Figure 16. Simulated Geant4 "net pit INS spectra" of Pit SiO_2 + 5%C for system with different shielding.

2.8.4. Dependence of "Soil Carbon Net Peak" area versus pit carbon content

The MC simulated and measured "pit net INS spectra" for pits with different carbon contents are shown in **Figure 12**. As can be seen, the simulated and measured spectra are similar. The simplicity of the model combined with not accounting for the detector efficiency with energy may help explain some differences between measured and simulated spectra. Despite the insignificant discrepancies between measured and simulated spectra, the main features (i.e., position and relative intensity of pair production; and silicon, oxygen, and carbon peaks) are approximately the same, and both were used in our analysis.

Assuming that the "soil carbon net peak" area value can be determined as "4.45 MeV net peak" area minus the "soil silicon net peak" area multiplied by some coefficient *f*, then the "4.45 MeV net peak" consists of only the "soil carbon net peak" and "silicon cascade transition peak," with the addition of other gamma rays being negligible. In this case, the dependence of the "soil carbon net peak" area versus carbon content should pass through the "zero-zero" point where the value of this coefficient equals the "cascade transition coefficient."

To define the dependence of the "soil carbon net peak" area versus carbon content for both the simulated and measured spectra, the "4.45 MeV net peak" area and "soil silicon net peak" area are determined in both the experimental and simulated spectra. In this case, spectral peaks of interest were approximated by one or two Gaussian shape curves using Igor Pro standard software [85] to determine the area beneath the curve. It is important to note that since peak fitting by summing two Gaussians gives approximately the same value for different component parameters, this sum was used in the analysis rather than the area of the components. An example of the simulated gamma spectra with fitted peaks with a centroid at 1.78 MeV ("soil silicon net peak") and 4.45 MeV ("4.45 MeV net peak") by Gaussian shape curves is shown in **Figure 17**.

The "soil carbon net peak area" in the *i*-th spectrum was denoted as $Ccorr_i$. The "soil silicon net peak" area in the *i*-th spectrum was denoted as SSi_i, "4.45 MeV net peak" area in the *i*-th spectrum as SC_i, and carbon content in the *i*-th mixture as $Cont_i$. The assumption was that $Ccorr_i$ can be calculated as $(SC_i - f{\cdot}SSi_i)$; $Ccorr_i$ was considered to be directly proportional to the

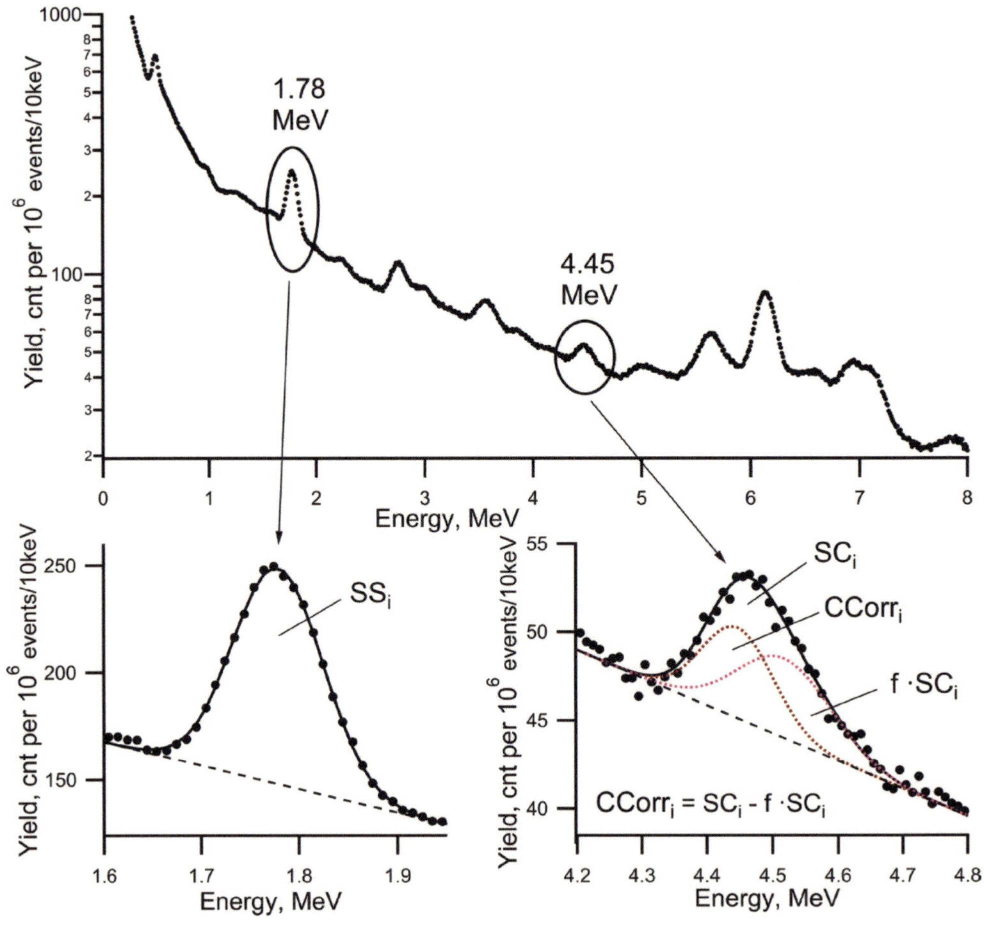

Figure 17. An example of the simulated gamma spectra for the model soil sample and designations of the peak areas used in the calculations: points—simulated data, solid lines—approximation by one (1.78 MeV, "net soil silicon peak") or sum of two Gaussians (4.45 MeV, "net 4.45 MeV peak"), dotted lines—peak components, and dashed lines—background.

carbon content of the mixture ($k{\cdot}Cont_i$) with f and k being the coefficients (these designations are shown in **Figure 17** for clarity). Using SC_i and SSi_i data, the values of f and k can be determined by minimizing the expression

$$\sum_i (SC_i - f \cdot SSi_i - k \cdot Cont_i)^2 \rightarrow min \tag{6}$$

The f and k values were found by equating the derivatives of this sum with respect to f and k set to zero. These calculations were performed using the standard mathematical software, MathCAD (Parametric Technology Corporation, 2013).

The dependencies between the "4.45 MeV net peak" area, "soil silicon net peak" area, and "soil carbon net peak" area *Ccorr* with carbon content from simulated and measured spectra are presented in **Figure 18**. As can be seen, the dependencies in both cases are similar to each other and pass through the "zero-zero" point. In addition, the values of the coefficient f from data processing of both the experimental and simulated spectra are very close (0.054 and 0.058, respectively). Values of this parameter (i.e., coefficient of the cascade transition for ^{28}Si nuclei) are similar to earlier published values [81, 86]. Thus, it is possible to define the "soil carbon net peak" area (from the "soil net INS spectra") as "4.45 MeV net peak" area minus "soil silicon net peak" area multiplied by some coefficient f, where f is the "cascade transition coefficient" equal

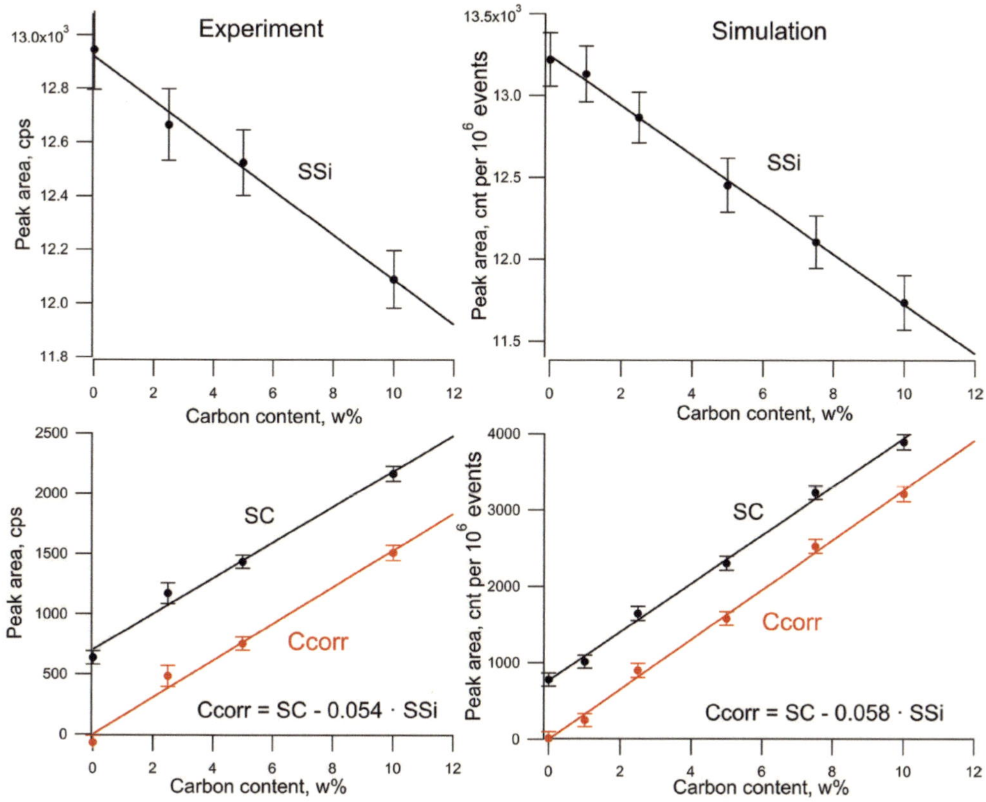

Figure 18. Dependencies of the "4.45 MeV net peak" area SC, "soil silicon net peak" area SSi, and "soil carbon net peak area" *Ccorr* with carbon content from the simulated and measured spectra of sand-carbon mixtures [79].

to 0.0547 without taking into account the effect of other INS processes like $^{16}O\ (n,n'\alpha)^{12}C^{*} \rightarrow {}^{12}C + \gamma$ (4.44 MeV).

2.9. Parameter selection for soil carbon characterization

The carbon gamma signal intensity ("soil carbon net peak" area) measured by the gamma detector is dependent on neutron flux intensity and soil conditions (density and element content), but the gamma signal intensity can be strongly influenced by the distribution of carbon within the soil depth profile. Neutron penetration depth and gamma flux attenuation are determined by soil properties. The distribution of soil carbon with depth is usually nonuniform (i.e., carbon level decreases as depth increases) and by first approximation can be described by exponential law [15]. The parameters of these distributions vary from site to site [56]. For this reason, correlations between the carbon peak intensity in the gamma spectrum and characterization of soil carbon content parameters are not obvious.

2.9.1. Parameter candidates

The main problem is determining which characteristic of soil carbon content has a direct proportional dependency (even in some approximation) with "soil carbon net peak" area. In general, the average carbon content or integral by some depth can be used to characterize the carbon in some depth layer. The tested candidates were average parameter—average carbon weight percent in some soil layer (*AvgCw%(h)*, where *h* is the layer thickness) and integral parameter—grams carbon per square centimeter of soil surface in a layer of some thickness (*SD(h)*, surface density). These parameters can be calculated as:

$$AvgCw\%(h) = \frac{1}{h}\int_0^h W\%(b)db \tag{7}$$

$$SD(h) = \int_0^h W\%(b) \cdot d(b)db \tag{8}$$

where *W%(b)* is carbon weight percent at depth *b* and *d(b)* is the soil density at depth *b*. Note that another possible characteristic will be proportional to one of these characteristics.

2.9.2. "Surface Density in 30 cm" parameter

Ref. [15] reported that the value of the carbon signal in INS spectra was connected to the surface density of carbon in a 30-cm layer. **Figure 19** shows three carbon depth profiles in modeled sand-carbon mixtures for which the values of *SD(30)* are all equal to 2.29 gC cm^{-2}. But the fragment of the MC simulated INS spectra around the carbon peak (**Figure 20**) for these modeled sand-carbon mixtures illustrates that these peaks are quite different despite *SD (30)* being the same for all mixtures. Thus, the "soil carbon net peak" area is not directly proportional to soil carbon content expressed in carbon surface density at 30 cm; this indicates that some other parameter should be found.

The effect of carbon depth profile and soil elemental content on the gamma spectrum as a whole (particularly for the carbon peak) can be determined from experimental results for soil sites with different carbon depth profiles, and by varying the carbon depth profile

http://dx.doi.org/10.5772/68014

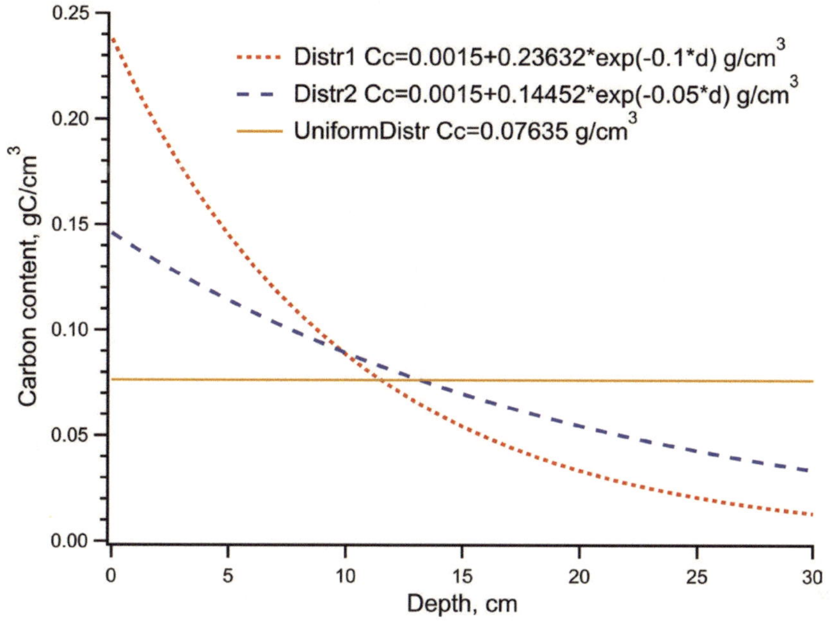

Figure 19. Carbon depth profiles in modeled sand-carbon mixtures for which the value of *SD(30)* = 2.29 gC cm^{-2}.

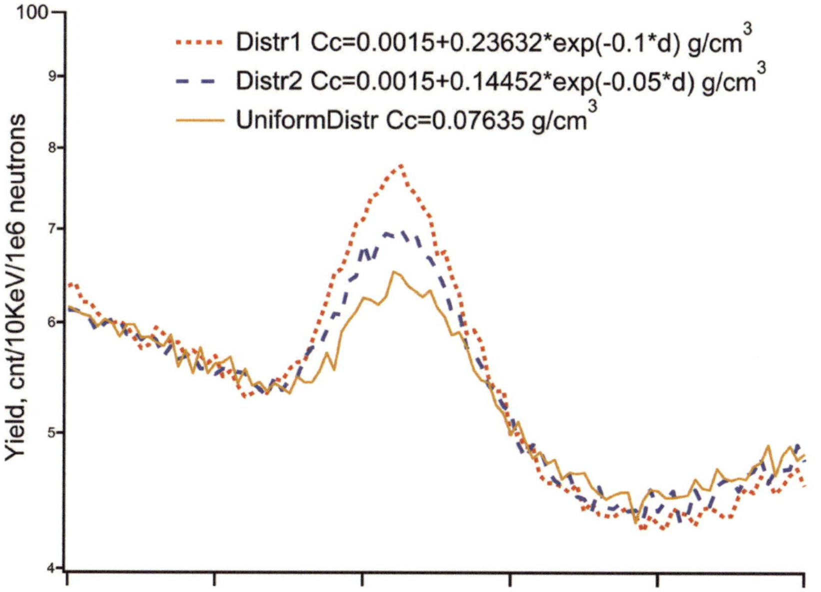

Figure 20. Fragment of MC simulated INS spectra (around the carbon peak) of modeled sand-carbon mixtures with carbon depth profiles shown in **Figure 19**. For all cases *SD(30)* = 2.29 gC cm^{-2}.

parameters in the soil model during MC simulations. These measurements and simulations were done to further our understanding of the relationship between INS signals and soil carbon content.

2.10. "Net INS Spectra" for nonuniform carbon depth profile sites

2.10.1. Carbon, soil density, and main element depth profile examples

Figures 21–23 show carbon depth profiles, soil density examples, and main element depth profiles from sampling sites. These carbon depth profiles were from an applied field (AF) located at the Piedmont Research Unit, Camp Hill, AL, USA [41]. Data was from traditional dry combustion chemical analysis of cores collected from the AF sites. Dependencies shown in **Figures 21–23** were used to construct the soil model in the simulation. Six artificial carbon depth profiles with extremal shapes (Art1–Art6 in **Figure 21**) were also used in the simulations.

2.10.2. Measured and simulated net INS spectra for sites with nonuniform carbon depth profile

Raw INS and TNC spectra were collected for each site and replotted in units of "counts per second." Afterwards, "soil net INS spectra" were calculated taking into account the "system background spectra" data (as was described above). For MC simulation, "soil net INS spectra" were calculated by subtracting the previously simulated system background spectra from the "soil INS spectra." Examples of measured and simulated spectra for some of these areas are

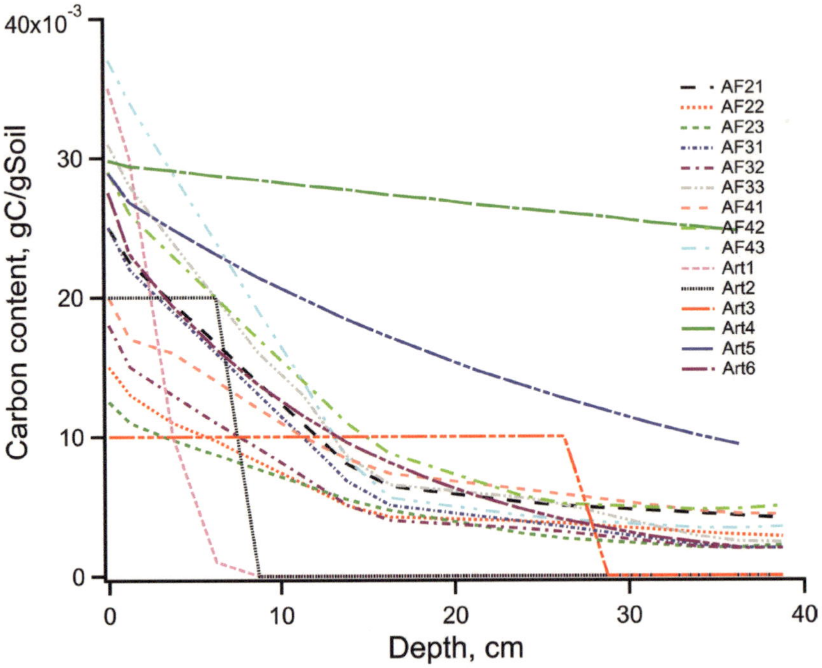

Figure 21. Soil carbon depth profiles for different sites (see text for details) [79].

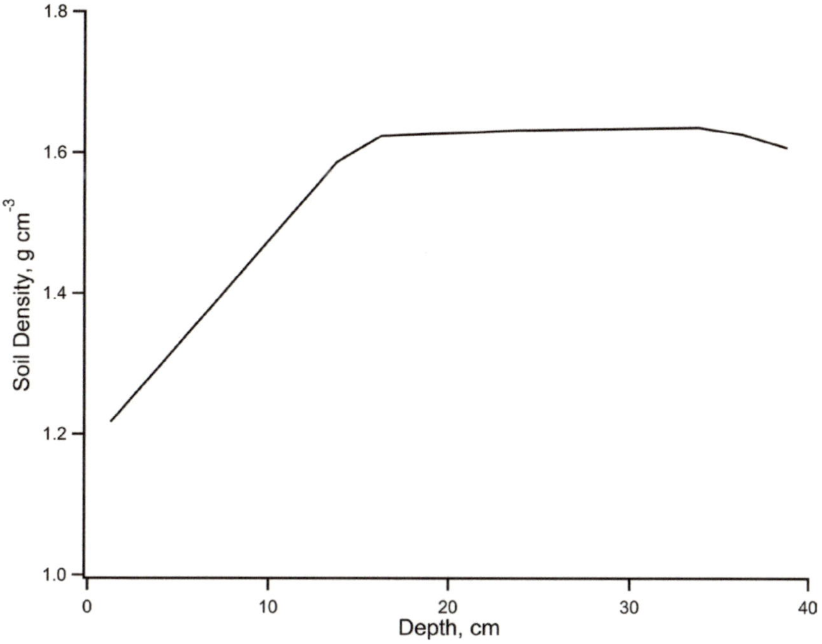

Figure 22. Soil density depth profile for the experimental site [79].

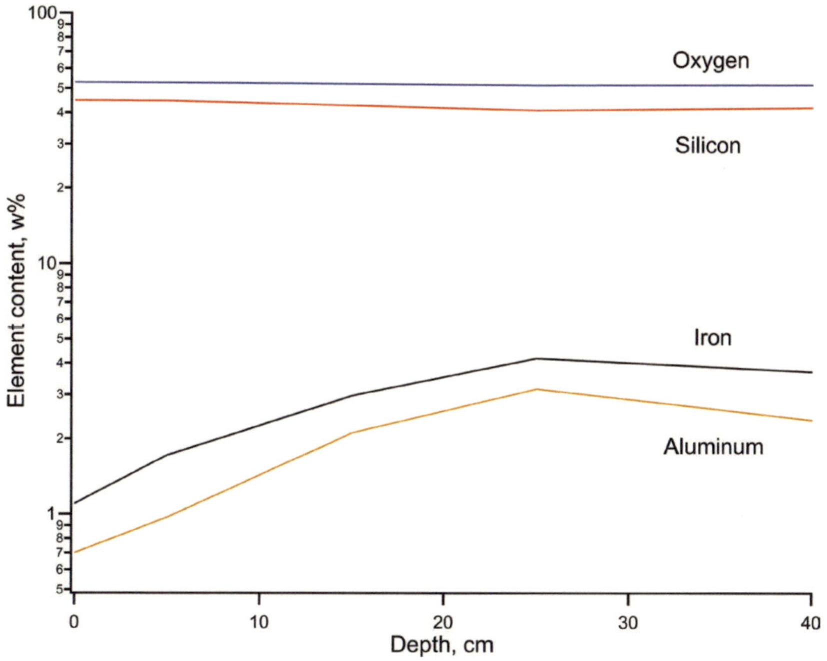

Figure 23. An example depth profile of the main soil elements [79].

shown in **Figure 24**. As can be observed, the simulated and measured spectra were very similar to each other. The peak areas with centroids at 1.78 and 4.45 MeV were calculated from these spectra using approximation by one or two Gaussian shape curves with Igor Pro standard software [85]. Next, "soil carbon net peak" area was defined using the above described procedure by subtracting the "soil silicon net peak" multiplied by the "cascade transition coefficient" (i.e., 0.0547) from the "4.45 MeV net peak" for each measurement and simulation.

2.10.3. Calibration

The calibration coefficient to calculate the carbon content from the value of "soil carbon net peak" area was determined from the gamma spectra for pits with uniform sand-carbon mixtures. Carbon content can be denoted in units of the average carbon weight percent or surface density. For uniform sand-carbon mixtures, the calibration line "soil carbon net peak" area versus $w\%$ will not depend on the thickness, while the calibration line "soil carbon net peak" area versus *SD* will depend on the given thickness. The carbon characterization parameter should be applied for any carbon depth profile, including uniform distribution. Thus, it should be possible to use the calibration coefficients derived from uniform distribution to calculate the carbon characterization parameter for the spectra of sites with nonuniform carbon depth distribution.

Calibration dependencies for uniform sand-carbon mixtures were also constructed for simulated and experimental spectra. In this case, the coefficients f and $k_{w\%,\,j}$ and $k_{SD,\,j}$ (j = 1 for measurement, j = 2 for MC simulation) were determined as described above by Eqs. (7) and (8) for both cases, where $Cont_i$ corresponded to $W\%$ or $SD(h)$. For uniform mixtures, $AvgCw\%$ does not depend on h. Thus, there is only one set of coefficients f and $k_{w\%,\,j}$. *SD* depends on h; therefore, each h has its own set of coefficients, f and $k_{SD,j}(h)$. In the set of coefficients for *SD*, the coefficient f is the same for each h and approximately the same for f for weight percent. Using the determined coefficients, the dependencies of the "soil carbon net peak" area (*Ccorr*) in the measured and MC simulated spectra versus carbon weight percent and versus carbon surface density at different thicknesses [$SD(h)$] in sand-carbon mixtures samples were plotted (**Figures 25** and **26**). These figures illustrate that the dependencies of *Ccorr* with $W\%$ and with $SD(h)$ are directly proportional within measurement and simulation accuracy limits in all cases.

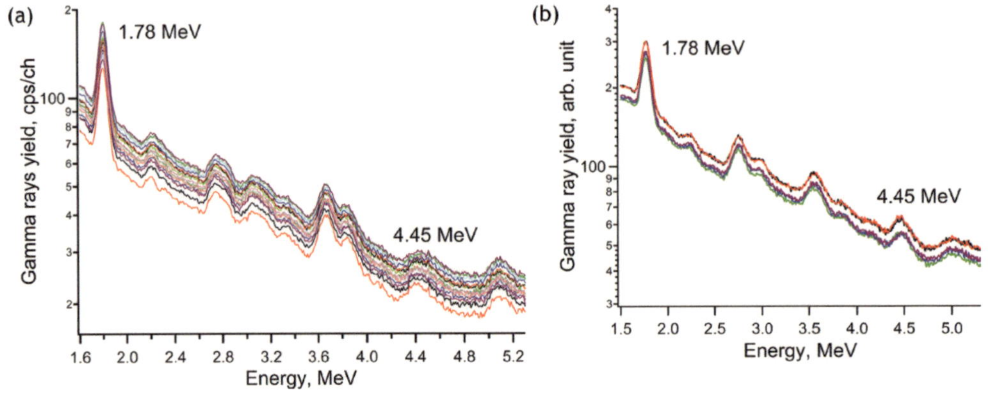

Figure 24. Measured (a) and MC simulated (b) "net soil INS gamma spectra" of different sites with characteristics shown in **Figures 21–23**.

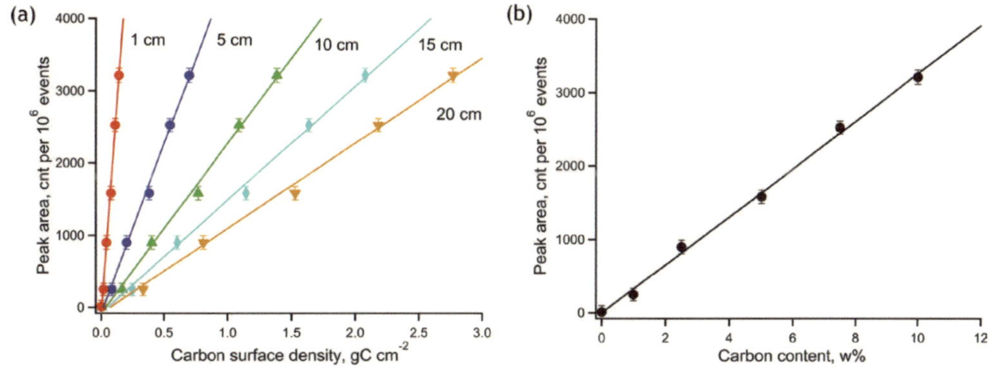

Figure 25. Calibration lines plotted from data of MC simulations [79]: (a) "net soil carbon peak" area versus carbon surface density; thickness shown near line; (b) "net soil carbon peak" area versus carbon weight percent.

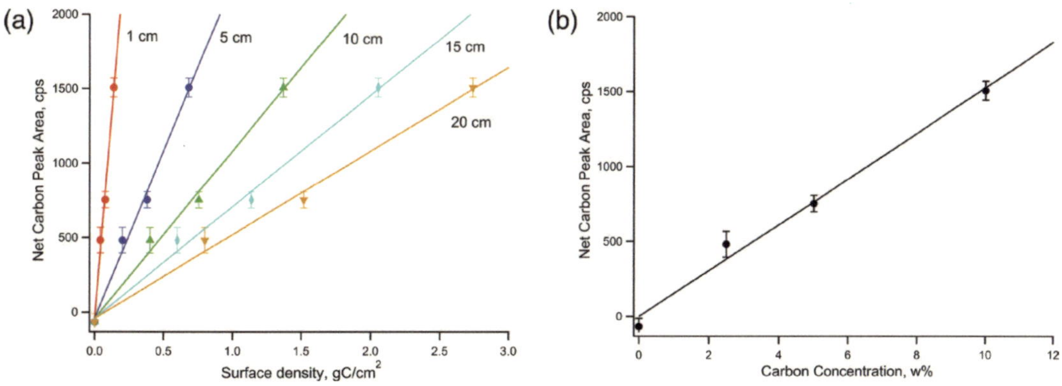

Figure 26. Calibration lines plotted from measurement data: (a) "net soil carbon peak" area versus carbon surface density; thickness shown near line; (b) "net soil carbon peak" area versus carbon weight percent [79].

2.11. Comparison of PFTNA and chemical analysis data

The calculation of $AvgCw\%(h)_{DC,i}$ and $SD(h)_{DC,i}$ by chemical analysis (dry combustion) was done for each site to compare with the data received from the INS gamma spectra. Coincidence of values for some parameters received from the net soil INS gamma spectra and from chemical analysis will mean the value of this parameter can be determined from neutron-gamma measurements.

2.11.1. *Dependence of average values of relative differences with* h

For each site, the relative difference between $Cw\%_{INS,i,j}$ and $SD_{INSi,j}(h)$ for INS and MC simulation data and $AvgCw\%(h)_{DC,i}$ and $SD(h)_{DC,i}$ values from soil chemical analysis data were used to compare these values.

$$rw_{\%i,j}(h) = \frac{AvgCw\%(h)_{DC,i} - Cw\%_{INS,i,j}}{Cw\%_{INS,i,j}} \tag{9}$$

$$rSD_{i,j}(h) = \frac{SD(h)_{DC,i} - SD_{INS,i,j}}{SD_{INS,i,j}} \tag{10}$$

The relative difference values for each site with h were calculated and plotted. The example of $rw_{\%i,2}(h)$ for different sites is shown in **Figure 27**. This relative difference was found to be equal to zero at some layer thickness h for each site. As can be seen in **Figure 27**, this depth varies around 10 cm in the range of ±2 cm for all sites.

The dependence of average values of relative differences for weight percent and for surface densities, $\xi w\%_j(h)$ and $\xi CD_j(h)$ for all surveyed sites with h, were calculated as

$$\xi w\%_j(h) = \frac{1}{N_j}\sum_i \frac{AvgCw\%(h)_{DC,i} - Cw\%_{INS,i,j}}{Cw\%_{INS,i,j}} \tag{11}$$

$$\xi CD_j(h) = \frac{1}{N_j}\sum_i \frac{SD(h)_{DC,i} - SD_{INS,i,j}}{SD_{INS,i,j}} \tag{12}$$

where N_j is the number of the sites used in measurements ($j = 1$) and in MC simulations ($j = 2$); both demonstrated some form of regularity. These dependencies are shown in **Figure 28** for both measurements and MC simulations.

Equality $\xi w\%_j(h)$ or $\xi CD_j(h)$ value to zero means that, at this h, the soil carbon characteristics determined from the spectra and from depth distribution are very similar. As one can see, the carbon weight percent derived from the spectra coincides with the average weight percent at a thickness of ~10 cm. **Figure 28** shows that the values of surface density from the spectra and from depth profiles differ from each other at any thickness. From these results we conclude

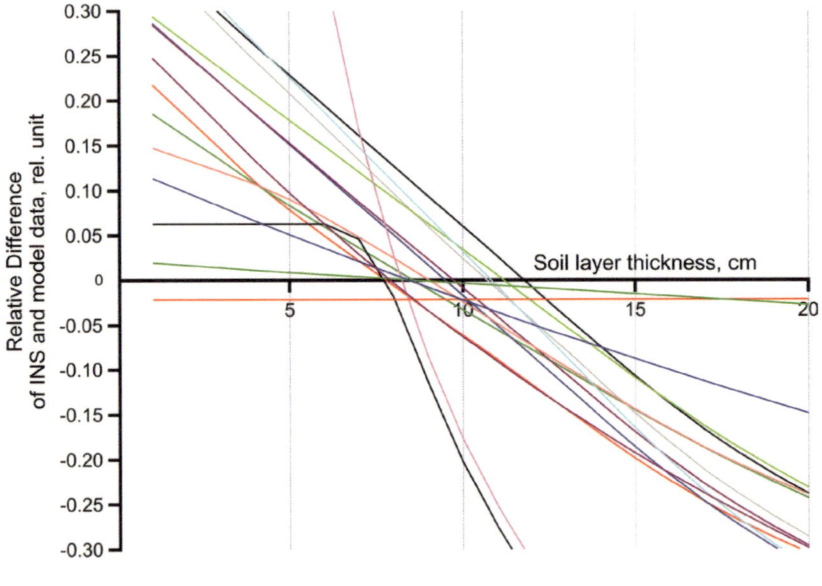

Figure 27. The dependence of the relative difference between $Cw\%_{INS,i,j}$ for data received from MC simulation and $AvgCw\%(h)_{DC,i}$ values from chemical analysis data with h for sites used for MC simulations [79].

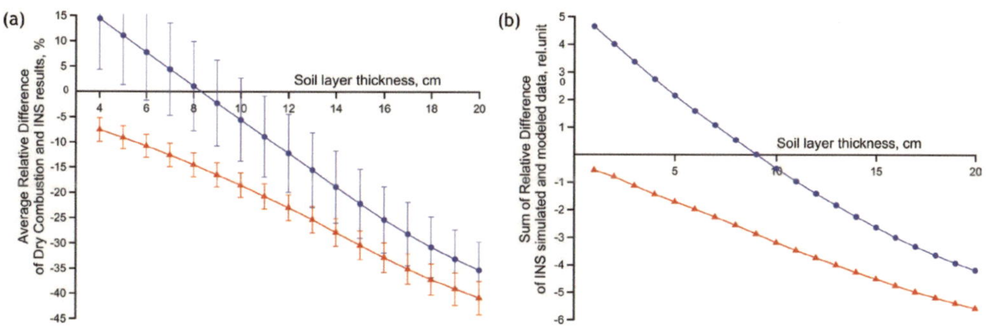

Figure 28. The dependence of average values of relative differences for weight percent (circles) and for surface densities (triangles) for measurements (a) and for MC simulations (b) with h [79].

that the soil carbon content parameter (based on gamma spectra using uniform carbon-sand mixture calibration data) is the average carbon weight percent for a 10-cm soil layer.

Therefore, INS simulation results (value of $Cw\%_{INS,i,2}$) can be attributed to average carbon weight percent in the soil layer with thickness h. Since different carbon depth profiles (from constant levels to sharp declines) were used in the simulations, the parameter (average carbon weight percent in soil layer with thickness 10 cm) could be assigned to the value determined from any INS gamma spectra.

2.11.2. *Average carbon weight percent measured by PFTNA and chemical analysis*

Results of carbon content measurements (average weight percent in upper 10-cm soil layer and its standard deviation) are shown in **Table 4**. Measurements were conducted by two methods (dry combustion and PFTNA). For clarity, the data from the open field at the Camp Hill location are shown in **Figure 29**. These data demonstrate good agreement between methods, especially for average values over whole plots. It should be noted that the accuracy of the carbon concentration measurement using PFTNA is comparable to measured values when the carbon concentration value is ~1 w% or less. To increase the accuracy of INS measurement at low soil carbon levels, further modification of our system is required. Such modifications would include (i) optimizing the detector's positioning relative to the neutron generator; (ii) increasing the number of detectors; and (iii) optimizing radiation shielding.

Data on soil carbon content can be used in mapping. Two maps of carbon distribution in the upper soil layer on one of the surveyed fields based on neutron-gamma analysis (PFTNA methods, **Figure 30a**) and chemical analysis (dry combustion, **Figure 30b**) are shown for comparison. As can be seen, both of these maps are very similar to each other. It should be noted that it took more than 1.5 months to collect the data for carbon content mapping in **Figure 30b** (dry combustion method), while only 2 working days were required for collecting carbon content data mapped in **Figure 30a** (PFTNA methods).

2.12. Effect of soil density and moisture on gamma response intensity

Soil density and moisture are parameters which could impact soil carbon measurement results when using PFTNA. While increasing soil density should increase the macroscopic

Location	Site # or Plot #	MINS measurements			Dry combustion measurements		
		Carbon, w%	STD, w%	Plot average ±STD, w%	Carbon, w%	STD, w%	Plot average ±STD, w%
Camp Hill Open Field	OF1	2.20	0.29	2.23±0.45	2.85	0.25	2.25±0.51
	OF2	2.51	0.29		2.54	0.31	
	OF3	1.76	0.22		1.91	0.13	
	OF4	1.88	0.23		2.99	0.94	
	OF5	2.82	0.25		3.03	0.37	
	OF6	2.15	0.21		1.99	0.26	
	OF7	2.77	0.32		1.92	0.41	
	OF8	2.52	0.25		2.44	0.15	
	OF9	2.06	0.26		1.79	0.27	
	OF10	2.17	0.27		2.25	0.45	
	OF11	2.39	0.22		2.23	0.30	
	OF12	3.11	0.31		2.91	0.47	
	OF13	1.44	0.25		1.49	0.42	
	OF14	1.93	0.29		1.80	0.19	
	OF15	1.86	0.27		1.67	0.25	
Camp Hill Applied Field 2	AF2-1	1.22	0.38	1.59±0.45	2.00	0.34	1.48±0.46
	AF2-2	2.09	0.37		1.14	0.34	
	AF2-3	1.46	0.37		1.31	0.08	
Camp Hill Applied Field 3	AF3-1	1.44	0.43	1.77±0.37	1.96	0.34	1.90±0.53
	AF3-2	1.68	0.37		1.34	0.34	
	AF3-3	2.17	0.39		2.4	0. 8	
Camp Hill Applied Field 4	AF4-1	2.59	0.42	2.33±0.34	1.58	0.34	2.12±0.46
	AF4-2	2.47	0.37		2.35	0.34	
	AF4-3	1.94	0.45		2.42	0.14	
E.V.Smith* Plots	S220	-	-	0.93±0.61	-	-	0.93±0.18
	S320	-	-	0.92±0.61	-	-	1.40±0.05
	S104	-	-	0.34±0.68	-	-	1.06±0.09
	S114	-	-	0.81±0.62	-	-	1.41±0.53
	S102	-	-	0.93±0.62	-	-	1.04±0.11
	S112	-	-	1.49±0.68	-	-	1.51±0.55

*The measurement was made for one site on these plots.

Table 4. Average carbon weight percent in the upper soil layer by dry combustion and PFTNA methods [79].

http://dx.doi.org/10.5772/68014

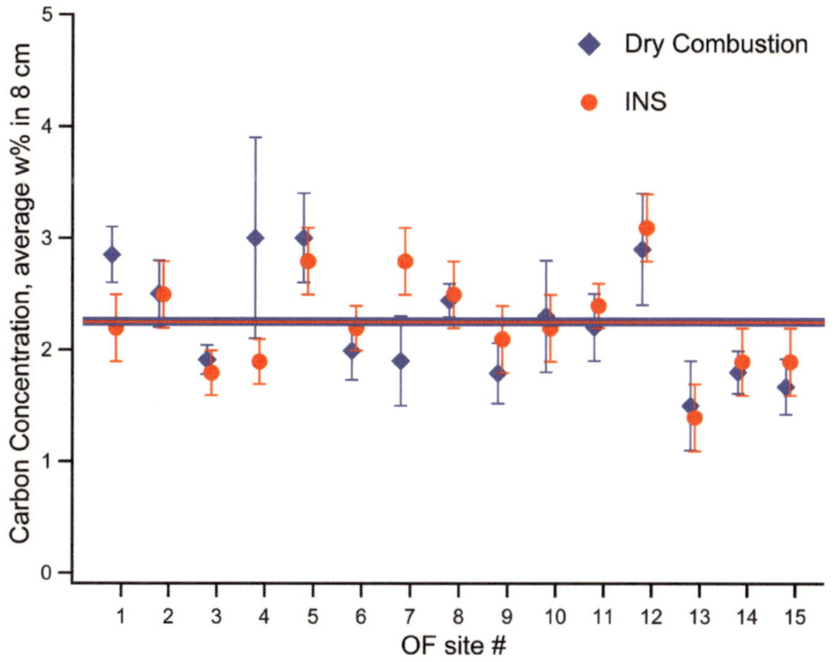

Figure 29. Average values of carbon weight percent for 10-cm soil layer measured by dry combustion (diamonds) and PFTNA (circles) methods for the open field (OF) site at Camp Hill (points) and average field values (solid lines) [79].

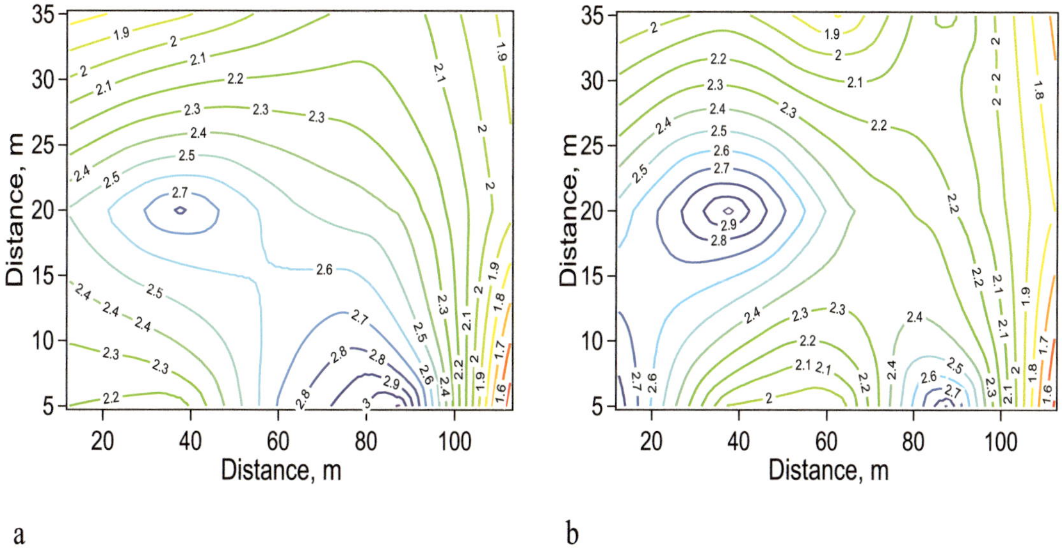

Figure 30. Carbon content maps of the upper layer of the open field (Piedmont Research Unit, Camp Hill, AL): (a) neutron-gamma analysis (PFTNA method) and (b) chemical analysis (dry combustion).

cross section of neutron interactions with soil nuclei, the excited soil volume could decrease. Thus, at first glance the effect of soil density on the peak of interest areas in the soil net INS spectra is not significant. The presence of soil moisture increases the amount of hydrogen atoms that run to faster neutron moderation and can decrease the peak of interest areas in the soil net INS spectra due to a decrease in fast neutron numbers. Using the Geant4 tool kit [82], MC simulations of gamma spectra for carbon-sand mixtures with different densities and moistures were conducted to estimate their effect on gamma response intensity. Both the INS and TNC spectra were simulated; the TNC processes were inactivated at INS spectra simulation (commands: /process/activate neutronInelastic, /process/inactivate nCapture, neutron data library JENDL4.0), while the INS processes were inactivated at TNC spectra simulation (commands: / process/activate nCapture, /process/inactivate neutronInelastic, neutron data library G4NDL4.5).

The simulated INS and TNC spectra for 150 cm × 150 cm × 60 cm pits with 5w% carbon-sand mixtures of different densities (from 1.1 to 1.52 g/cm^3) are shown in **Figure 31**. The dependencies of peak of interest areas with centroids at 1.78 and 4.45 MeV in the INS spectra with densities are shown in **Figure 32**. As can be seen in these figures, there are no significant changes in the spectra or peak areas. Thus, there is no significant effect of soil density on the INS spectra.

Simulated INS and TNC spectra for 150 cm × 150 cm × 60 cm pits with 5w% carbon-sand mixtures having different moistures *H* (*H* from 0 to 30%; a real range of soil moisture change) are shown in **Figure 33**. The dependencies of peak of interest areas with centroids at 1.78 and 4.45 MeV in the INS spectra with water weight percent *W* [*W*=*H* / (1+*H*)] are shown in **Figure 34**. As seen in these figures, the peak areas slightly decrease throughout

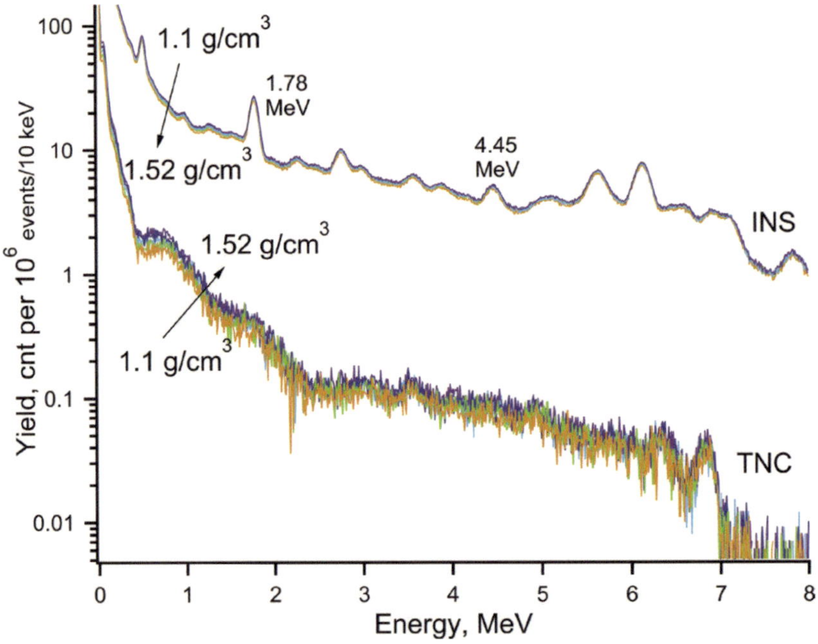

Figure 31. Geant4 simulated INS (JENDL4.0, 1e9 events) and TNC (G4NDL4.5, 1e9 events) gamma spectra for 150 cm × 150 cm × 60 cm pits with 5w% carbon-sand mixtures with different densities (from 1.1 to 1.52 g/cm^3) irradiated by 14.1 MeV neutrons.

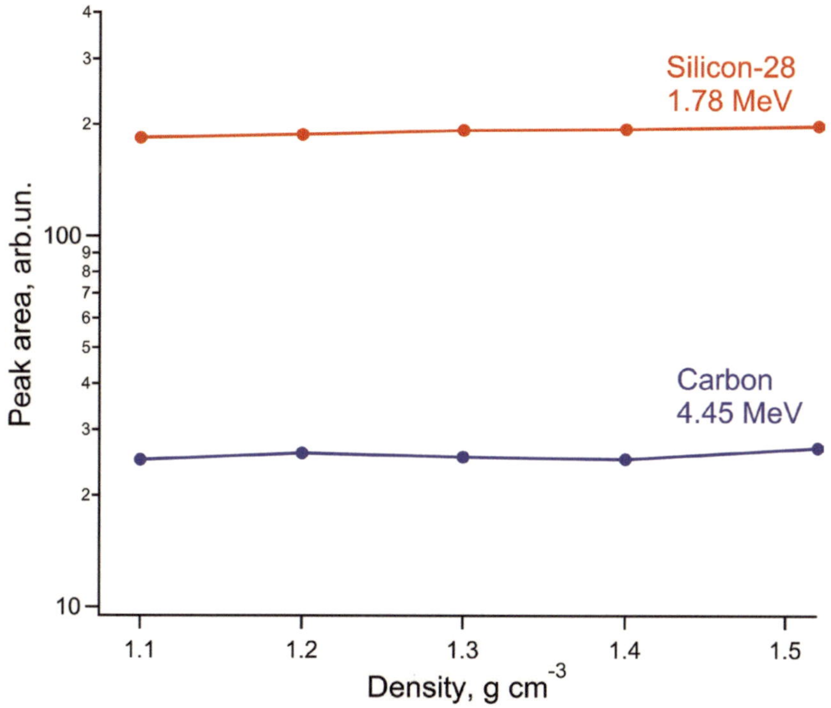

Figure 32. Dependencies of peak areas with centroids at 1.78 and 4.45 MeV with different densities in INS spectra shown in Figure 31.

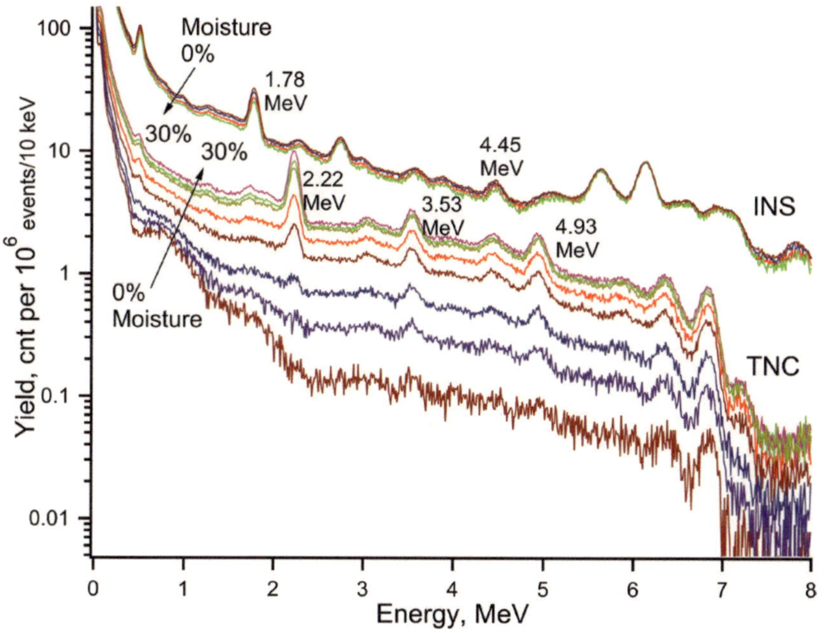

Figure 33. Geant4 simulated INS (JENDL4.0, 1e9 events) and TNC (G4NDL4.5, 1e9 events) gamma spectra for 150 cm × 150 cm × 60 cm pits with 5w% carbon-sand mixtures with different moistures (from 0 to 30%) irradiated by 14.1 MeV neutrons.

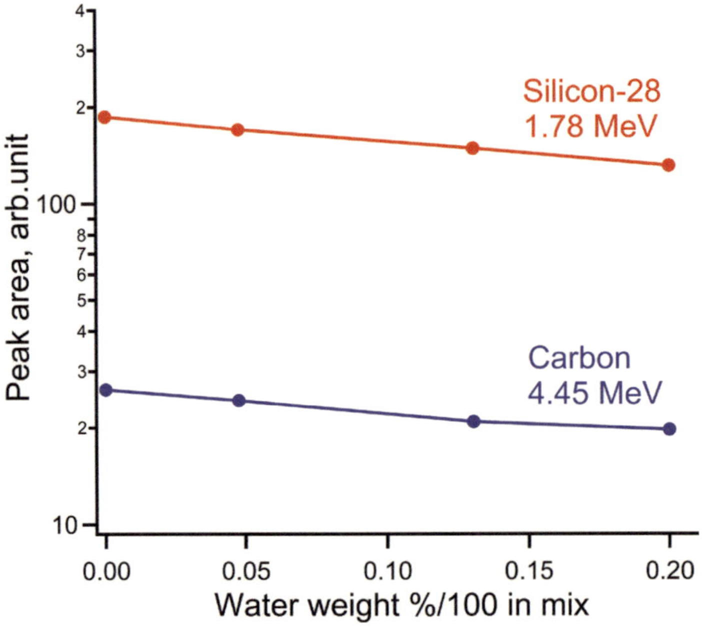

Figure 34. Dependencies of peak areas with centroids at 1.78 and 4.45 MeV with water weight percent in the INS spectra shown in **Figure 33**.

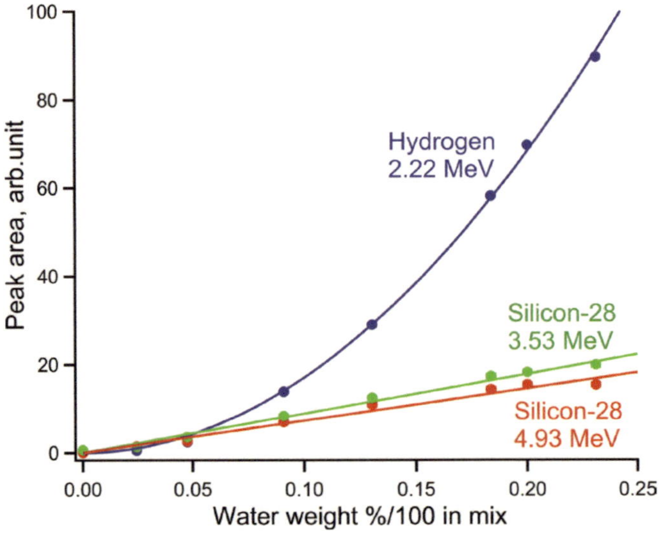

Figure 35. Dependencies of peak areas with centroids at 2.22, 3.53, and 4.93 MeV with water weight percent in the TNC spectra shown in **Figure 33**.

the possible moisture range. However, these changes are not significant in the practical range of soil moisture. Thus, it was concluded that moisture has no significant effect on the INS spectra.

At the same time, TNC spectra of gamma response increase with increasing soil moisture. In the studied moisture range, some peaks increase in direct proportion to the water weight percent in soil (e.g., TNC Si peaks with centroids at 3.53 and 4.93 MeV). The hydrogen peak with a centroid at 2.22 MeV increases as a square of the water weight percent within error limits (**Figure 35**). This probably occurs due to the linear increase of both thermal neutron flux and number of hydrogen nuclei as water weight percent increases. The conclusions regarding moisture and density effects on soil INS gamma response spectra agree with findings of others [15, 87].

3. Conclusion

Results of the PFTNA and the "gold standard" chemical analysis (Dry Combustion Technique) demonstrated good agreement for soil carbon content measurements in the upper soil layer (~10 cm). Experimental results successfully demonstrated that the average carbon weight percent in the upper soil layer (regardless of carbon depth distribution shape) can be measured *in situ* by the PFTNA measurement method (1 h) with accuracy comparable to the "gold standard" technique. The described procedures for background accountability, system calibration, and "soil carbon net peak" area calculations from the acquired spectra should be utilized. Although the current mobile system for PFTNA is fully capable of routine soil carbon measurements in natural and agricultural field settings, future modifications of the detector system and shielding can improve measurement accuracy and decrease measurement time. Nevertheless, the main features and herein described procedures (i.e., system background determination, calibration procedure, and "soil carbon net peak" area extraction) indicate that PFTNA methods can be recommended as a viable alternative procedure for soil carbon measurement. Additionally, MC simulations showed that soil density and moisture do not significantly impact soil carbon measurements by PFTNA.

Acknowledgements

The authors are indebted to Barry G. Dorman, Robert A. Icenogle, Juan Rodriguez, Morris G. Welch, and Marlin Siegford for technical assistance in experimental measurements, and to Jim Clark and Dexter LaGrand for assistance with computer simulations. We thank XIA LLC for allowing the use of their electronics and detectors in this project. This work was supported by NIFA ALA Research Contract No ALA061-4-15014 "Precision geospatial mapping of soil carbon content for agricultural productivity and lifecycle management."

Author details

Aleksandr Kavetskiy, Galina Yakubova, Stephen A. Prior and Henry Allen Torbert*

*Address all correspondence to: allen.torbert@ars.usda.gov

USDA-ARS National Soil Dynamics Laboratory, Auburn, Alabama, USA

References

[1] Valkovic V. 14 MeV Neutrons: Physics and Applications. CRC Press: Taylor & Francis Group; 2016, 481 p. DOI:10.1063/1.3120103

[2] Aleksahkin V, Bystritsky V, Zamyatin N, Zubarev E, Krasnoperov A, Rapatsky V, Rogov YU, Sadovsky A, Salamatin A, Salmin R, Sapozhnikov M, Slepnev V, Khabarov S, Razinkov E. Detection of diamonds in kimberlite by the tagged neutron method. Nuclear Instruments and Methods in Physics Research A. 2015; **785**:9–13. DOI:10.1016/j.nima.2015.02.049

[3] Parsons A, Bodnarik J, Evans L, Floyd S, Lim L, McClanahan T, Namkung M, Nowicki S, Schweitzer J, Starr R, Trombka J. Active neutron and gamma-ray instrumentation for in situ planetary science applications. Nuclear Instruments and Methods in Physics Research A. 2011; **652**:674–679. DOI:10.1016/j.nima.2010.09.157

[4] Miceli A, Festa G, Gorini G, Senesi R, Andreani C. Pulsed neutron gamma-ray logging in archaeological site survey. Measurement Science and Technology. 2013; **24**:125903–125908. DOI:10.1088/0957-0233/24/12/125903

[5] Tykot R. Scientific methods and applications to archaeological provenance studies. In: Physics Methods in Archaeometry. Proceedings of the International School of Physics "Enrico Fermi". Martini M, Milazzo M, Piacentini M, editors. Bologna, Italy: Società Italiana di Fisica; 2014, pp. 407–432.

[6] Panjech H, Izadi-Najafabadi R. Body composition analyzer based on PGNAA method. In: Radioisotopes—Applications in Physical Sciences. Nirmal S, editor. InTech; 2011, pp. 311–324. ISBN 978-953-307-510-5.

[7] Chichester D, Empey E. Measurement of nitrogen in the body using a commercial PGNAA system—phantom experiments. Applied Radiation and Isotopes. 2004; **60**:55–61. DOI:10.1016/j.apradiso.2003.09.007

[8] Mitra S, Wolff J, Garrett R, Peters C. Application of the associated particle technique for the whole-body measurement of protein, fat and water by 14 MeV neutron activation analysis—a feasibility study. Physics in Medicine and Biology. 1995; **40**:1045–1055.

[9] Mitra S, Wolff J, Garrett R. Calibration of a prototype *in vivo* total body composition analyzer using 14 MeV neutron activation and the associated particle technique. Applied Radiation and Isotopes. 1998; **49**:537–539.

[10] Thermo Fisher Scientific Inc. 2016. Available from: http://www.thermofisher.com/search/browse/results?customGroup=Neutron+Generators [Accessed: 2017-01-26]

[11] Dep L, Belbot M, Vourvopoulos G, Sudar S. Pulsed neutron-based on-line coal analysis. Journal Radioanalytical Nuclear Chemistry. 1998; **234**:107–112. DOI:10.1007/BF02389756

[12] Saleh H, Livingston R. Experimental evaluation of a portable neutron-based gamma-spectroscopy system for chloride measurements in reinforced concrete. Journal of Radio-analytical and Nuclear Chemistry. 2000; **244**:367–371. DOI:10.1023/A:1006787626016

[13] Naqvi A, Kalakada Z, Al-Matouq F, Maslehuddin M. Prompt gamma-ray analysis of chlorine in superpozz cement concrete. Nuclear Instruments and Methods in Physics Research A. 2012; **693**:67–73. DOI:10.1016/j.nima.2012.06.059

[14] Nikitin A, Bliven S. Needs of well logging industry in new nuclear detectors. Nuclear Science Symposium Proceedings. 2010; 1214–1219. DOI:10.1109/NSSMIC.2010.5873961

[15] Wielopolski L. Nuclear methodology for non-destructive multi-elemental analysis of large volumes of soil. In: Planet Earth: Global Warming Challenges and Opportunities for Policy and Practice. Carayannis E, editor. 2011, pp. 467–492. DOI:10.5772/23230

[16] Seybold C, Mausbach M, Karlen D, Rogers H. Quantification of soil quality. Soil Processes and the Carbon Cycle. In: Lal R, Kimble J, Stewart B, editors. Boca Raton; FL: CRC Press; 1997, pp. 387–404

[17] Potter K, Daniel J, Altom W, Torbert H. Stocking rate effect on soil carbon and nitrogen in degraded soils. Journal of Soil Water Conservation. 2001; **56**:233–236

[18] Torbert H, Prior S, Runion G. Impact of the return to cultivation on carbon (C) sequestration. Journal of Soil Water Conservation. 2004; **59**:1–8

[19] Stolbovoy V, Montanarella L, Filippi N, Jones A, Gallego J, Grassi G. Soil sampling protocol to certify the changes of organic carbon stock in mineral soil of the European Union. Version 2. EUR 21576 EN/2. Luxembourg: Office for Official Publications of the European Communities; 2007, 56 pp. ISBN 978-92-79-05379-5

[20] Smith K, Watts D, Way T, Torbert H, Prior S. Impact of tillage and fertilizer application method on gas emissions (CO_2, CH_4, N_2O) in a corn cropping system. Pedosphere. 2012; **22**:604–615

[21] Gilmore G. Practical Gamma-Ray Spectrometry. 2nd ed. West Sussex, England: Wiley; 2008, 387 p

[22] NNDC. Upton, NY: Brookhaven National Laboratory; 2013. Available at: http://www.nndc.bnl.gov/chart [Accessed: 2017-01-26]

[23] Yang J, Yang Y, Li Y, Tuo X, Li Z, Cheng Y, Mou Y, Huang W. Prompt gamma neutron activation analysis for multi-element measurement with series samples. Laser Physics Letters. 2013; **10**:1–5. DOI:10.1088/1612-2011/10/5/056002

[24] Mernagh J, Harrison J, McNeill K. *In vito* determination of nitrogen using Pu-Be sources. Physics in Medicine and Biology. 1977; **22**:831–835.

[25] Shue S, Faw R, Shultis J. Thermal-neutron intensities in soils irradiated by fast neutrons from point sources. Chemical Geology. 1998; **144**:47–61. DOI:10.1016/S0009-2541(97)00108-3

[26] Chung C, Tseng T-C. *In situ* prompt gamma-ray activation analysis of water pollutants using a shallow ^{252}Cf-HPGe Probe. Nuclear Instruments and Methods in Physics Research Section A. 1988; **267**:223–230. DOI:10.1016/0168-9002(88)90651-1

[27] Barzilov A, Novikov I, Womble P. Material analysis using characteristic gamma rays induced by neutrons, gamma radiation. In: Gamma Radiation. Adrovic F, editor. InTech; 2012. Available from: http://www.intechopen.com/books/gamma-radiation/material-analysis-usingcharacteristic-gamma-rays-induced-by-pulse-neutrons. DOI:10.5772/2054

[28] NRC. 2010. Available from: http://www.nrc.gov/docs/ML1122/ML11229A704.pdf [Accessed: 2017-01-26]

[29] Industrial Radiation Sources. Eckert & Ziegler. 2008. Available from: http://www.ezag.com/fileadmin/ezag/user-uploads/isotopes/isotopes/5_industrial_sources.pdf [Accessed: 2017-01-26]

[30] HPS (Health Physics Society). 2015. Available from: https://hps.org/publicinformation/ate/q8216.html [Accessed: 2017-01-26]

[31] FTC (Frontier Technology Corporation). 2015. Available from: http://www.frontier-cf252.com/index.html [Accessed: 2017-01-26]

[32] Sodern. 2015. http://www.sodern.com/sites/en/ref/Neutron-generator_79.html [Accessed: 2017-01-26]

[33] Thermo Fisher Scientific Inc. 2016. Available from: https://www.thermofisher.com/order/catalog/product/19187 [Accessed: 2017-01-26]

[34] VNIIA. 2015. Available from: http://vniia.ru/eng/ng/element.html [Accessed: 2017-01-26]

[35] Saint-Gobain. Scintillation products. Technical Note. 2009. Available from: http://www.crystals.saint-gobain.com/sites/imdf.crystals.com/files/documents/brillance_performance_summary.pdf [Accessed: 2017-01-26]

[36] Feng X, Cheng C, Yin Z, Khamlary M, Townsend P. Two kinds of deep hole-traps in $Bi_4Ge_3O_{12}$ crystals. Chinese Physics Letters. 1992; **9**:597–600.

[37] Naqvi A, Kalakada Z, Al-Matouq F, Maslehuddin M, Al-Amoudi O. Chlorine detection in fly ash concrete using a portable neutron generator. Applied Radiation and Isotopes. 2012; **70**:1671–1674.

[38] Van Loef E, Dorenbos P, van Eijk C, Krämer K, Güdel H. High-energy-resolution scintillator: Ce^{3+} activated $LaBr_3$. Applied Physics Letters. 2001; **79**:1573–1575. DOI:10.1063/1.1385342

[39] McFee J, Faust A, Andrews H, Kovaltchouk V, Clifford E, Ing H. A comparison of fast inorganic scintillators for thermal neutron analysis landmine detection. IEEE Transactions on Nuclear Science. 2009; **56**:1584–1592. DOI:10.1109/TNS.2009.2018558

[40] Turner J. Atom, Radiation and Radiation Protection. 3rd ed. Weinheim, Germany: Willey-VCH Verlag GmbH & Co. KGaA; 2007, 586 p. ISBN 978-3-527-40606-7

[41] Kavetskiy A, Yakubova G, Torbert H, Prior S. Application of Geant4 simulation for analysis of soil carbon inelastic neutron scattering measurements. Applied Radiation and Isotopes. 2016; **113**:33–39. DOI:10.1016/j.apradiso.2016.04.013

[42] Sudac D, Valkovic V. Irradiation of 4"x4" NaI(Tl) detector by the 14 MeV neutrons. Applied Radiation and Isotopes. 2010; **68**:896–900. DOI:10.1016/j.apradiso.2009.12.014

[43] IAEA. Neutron generators for analytical purposes. 2012, 162 p. ISSN 2225-8833(1). Available from: http://www-pub.iaea.org/MTCD/Publications/PDF/P1535_web.pdf [Accessed: 2017-01-26]

[44] Saint-Gobain. Efficiency calculations for selected scintillators. Available from: http://www.crystals.saint-gobain.com/uploadedFiles/SG-Crystals/Documents/Technical/SGC%20Efficiency%20Calculations%20Brochure.pdf [Accessed: 2017-01-26]

[45] Naqvi A, Maslehuddin M, Kalakada Z, Al-Amoudi O. Prompt gamma ray evaluation for chlorine analysis in blended cement concrete. Applied Radiation and Isotopes. 2014; **94**:8–13. DOI:10.1016/j.apradiso.2014.06.011

[46] Womble P, Vourvopoulos G, Paschal J, Novikov I, Barzilov A. Results of field trials for the PELAN system. Proceedings of SPIE. 2002; **4786**: 52–57. DOI:10.1117/12.450506

[47] Bystritsky V, Zubarev E, Krasnoperov A. Porohovoi S, Rapatskii V, Rogov Yu, Sadovskii A, Salamatin A, Salmin R, Slepnev V, Andreev E. Gamma detectors in explosives and narcotics detection systems. Physics of Particles and Nuclei Letters. 2013; **10**: 566–572. DOI:10.1134/S154747711306006X

[48] Ozdemir T, Akbay I, Uzun H, Reyhancan I. Neutron shielding of EPDM rubber with boric acid: mechanical, thermal properties and neutron absorption tests. Progress in Nuclear Energy. 2016; **89**:102–109. DOI:10.1016/j.pnucene.2016.02.007

[49] Elmahroug Y, Tellili B, Souga C. Calculation of gamma and neutron shielding parameters for some materials polyethylene-based. International Journal of Physics and Research. 2013; **3**:33–40

[50] Singh V, Badiger N. Investigation of gamma and neutron shielding parameters for borate glasses containing NiO and PbO. Physics Research International. 2014. Available from: https://www.hindawi.com/journals/physri/2014/954958/ [Accessed: 2017-01-26]. DOI:10.1155/2014/954958

[51] Elsheikh N, Habbani F, ElAgib I. Investigations of shield effect and type of soil on landmine detection. Nuclear Instruments and Methods in Physics Research A. 2011; **652**:1–4. DOI:10.1016/j.nima.2010.09.183

[52] El-Bakkoush F, Akki T. Attenuation of neutrons and total gamma rays in two layers shields. Journal of Nuclear Science and Technology. 2000; **37**:526–529. DOI:10.1080/00223131.2000.10874943

[53] Wielopolski L, Mitra S, Doron O. Non-carbon-based compact shadow shielding for 14 MeV neutrons. Journal of Radioanalysis Nuclear Chemistry. 2008; **276**:179–182. DOI:10.1007/s10967-007-0429-1

[54] Bystritsky V, Valkovich V, Grozdanov D, Zontikov A, Ivanov I, Kopach Yu, Krulov A, Rogov Yu, Ruskov I, Sapoznikov M, Skoii V, Shvezov V. Multilayer passive shielding of the scintillation detectors on the base of BGO, NaI(Tl) and stilbene working in the intensive neutron fields with energy 14.1 MeV. Physics of Particles and Nuclei Letters. 2015; **12**:325–335. DOI:10.1134/S1547477115020089

[55] Mitra S, Dioszedi I. Development of an instrument for non-destructive identification of unexploded ordnance using tagged neutrons—a proof of concept study. IEEE Nuclear Science Symposium Conference Record. 2011; 285–289. DOI:10.1109/NSSMIC.2011.6154499

[56] Yakubova G, Wielopolski L, Kavetskiy A, Torbert H, Prior S. Field testing a mobile inelastic neutron scattering system to measure soil carbon. Soil Science. 2014; **179**:529–535. DOI:10.1097/SS.0000000000000099

[57] Womble P, Schultz F, Vourvopoulos G. Non-destructive characterization using pulsed fast-thermal neutrons. Nuclear Instrument and Methods in Physics Research Section B. 1995; **99**:757–760. DOI:10.1016/0168-583X(95)00326-6

[58] Nellis W. Slowing-down distances and times of 0.1–14 MeV neutrons in hydrogenous materials. American Journal of Physics. 1977; **45**:443–446. DOI:10.1119/1.10833

[59] Miller V. Measuring the mean lifetime of thermal neutrons from a small specimen. Atomnaya Energiya. 1967; **22**:33–38. DOI:10.1007/BF01225391

[60] Vourvopoulos G, Womble P. Pulsed fast/thermal neutron analysis: a technique for explosives detection. Talanta. 2001; **54**:459–468. DOI:10.1016/S0039-9140(00)00544-0

[61] Mitra S, Wielopolski L. Optimizing the gate-pulse width for fast neutron induced gamma-ray spectroscopy. Proceedings of SPIE. 2005; **5923**(8):16. DOI:10.1117/12.614569

[62] Nelson D, Sommers L. Total carbon, organic carbon, and organic matter. In: Sparks D, editor. Methods of Soil Analysis. Madison, WI: SSSA and ASA; 1996, pp. 961–1010.

[63] Hendricks J. A Monte Carlo code for particle transport. Los Alamos Science. 1994; **22**:42–33.

[64] Lux L, Koblinger L. Monte Carlo particle transport methods: neutron and photon calculations. Florida: CRC Press; 1991, 517 p

[65] Deiev O. Geant4 simulation of neutron transport and scattering in media. Problems of atomic science and technology. Series: Nuclear Physics Investigation. 2013; **3**:236–241

[66] Bak S, Park T-S, Hong S. Geant4 simulation of the shielding of neutrons from ^{252}Cf source. Journal of the Korean Physical Society. 2011; **59**:2071–2074. DOI:10.3938/jkps.59.2071

[67] Reda A. Monte Carlo simulations of a D-T neutron generator shielding for landmine detection. Radiation Measurements. 2011; **46**:1187–1193. DOI:10.1016/j.radmeas.2011.07.013

[68] Feng Z, Jun-tao L. Monte Carlo simulation of PGNAA system for determining element content in the rock sample. Journal of Radioanalytical and Nuclear Chemistry. 2014; **299**:1219–1224. DOI:10.1007/s10967-013-2858-3

[69] Qin X, Zhou R, Han J-F, Yang C-W. GEANT4 simulation of the characteristic gamma-ray spectrum of TNT under soil induced by DT neutrons. Nuclear Science and Techniques. 2015; **26**:010501-1–010501–6. DOI:10.13538/j.1001-8042/nst.26.010501

[70] Nasrabadi M, Bakhshi F, Jalali M, Mohammadi A. Development of a technique using MCNPX code for determination of nitrogen content of explosive materials using prompt gamma neutron activation analysis method. Nuclear Instruments and Methods in Physics Research Section A. 2011; **659**:378–382. DOI:10.1016/j.nima.2011.08.029

[71] Perot B, El Kanawati W, Carasco C, Eleon C, Valkovic V, Sudac D, Obhodas J, Sannie, G. Quantitative comparison between experimental and simulated gamma-ray spectra induced by 14 MeV tagged neutrons. Applied Radiation and Isotopes. 2012; **70**:1186–1192. DOI:10.1016/j.apradiso.2011.07.005

[72] El Kanawati W, Perot B, Carasco C, Eleon C, Valkovic V, Sudac D, Obhodas J, Sannie G. Acquisition of prompt gamma-ray spectra induced by 14 MeV neutrons and comparison with Monte Carlo simulations. Applied Radiation and Isotopes. 2011; **69**:732–743. DOI:10.1016/j.apradiso.2011.01.010

[73] Tain J, Agramunt J, Algora A, Aprahamian A, Cano-Ott D, Fraile L, Guerrero C, Jordan M, Mach H, Martinez T, Mendoza, Mosconi M, Nolte R. The sensitivity of $LaBr_3$:Ce scintillation detectors to low energy neutrons: measurement and Monte Carlo simulation. Nuclear Instruments and Methods in Physics Research Section A. 2015; **774**:17–24. DOI:10.1016/j.nima.2014.11.060

[74] Wielopolski L, Song Z, Orion I, Hanson A, Hendrey G. Basic considerations for Monte Carlo calculations in soil. Applied Radiation and Isotopes. 2005; **62**:97–107. DOI:10.1016/j.apradiso.2004.06.003

[75] Doron O, Wielopolski L, Mitra S, Biegalski S. MCNP benchmarking of an inelastic neutron scattering system for soil carbon analysis. Nuclear Instruments and Methods in Physics Research Section A. 2014; **735**:431–436. DOI:10.1016/j.nima.2013.09.049

[76] Wielopolski L, Yanai R, Levine C, Mitra S, Vadeboncoeur M. Rapid, non-destructive carbon analysis of forest soils using neutron-induced gamma-ray spectroscopy. Forest Ecology Management. 2010; **260**:1132–1137. DOI:10.1016/j.foreco.2010.06.039

[77] Mitra S, Wielopolski L, Tan H, Fallu-Labruyere A, Hennig W, Warburton W. Concurrent measurement of individual gamma-ray spectra during and between fast neutron pulses. Nuclear Science. 2007; **54**:192–196. DOI:10.1109/TNS.2006.889165

[78] Tan H, Mitra S, Wielopolski L, Fallu-Labruyere A, Hennig W, Chu Y, Warburton W. A multiple time-gated system for pulsed digital gamma-ray spectroscopy. Journal of Radioanalysis Nuclear Chemistry. 2008; **276**:639–643. DOI:10.1007/s10967-008-0611-0

[79] Yakubova G, Kavetskiy A, Prior S, Torbert H. Benchmarking the inelastic neutron scattering soil carbon method. Vadose Zone Journal. 2016; **15**:1–11. DOI:10.2136/vzj2015.04.0056.

[80] Murphy R, Kozlovsky B, Share G. Nuclear cross sections for gamma-ray de-excitation line production by secondary neutrons in the Earth's atmosphere. Journal of Geophysics Research. 2011; **116**:1–9. DOI:10.1029/2010JA015820

[81] Herman M, Capote R, Carlson B, Oblozinsky P, Sin M, Trkov A, Wienke H, Zerkin V. EMPIRE: nuclear reaction model code system for data evaluation. Nuclear Data Sheets. 2007; **108**:2655–2715. DOI:10.1016/j.nds.2007.11.003

[82] Agostinelli S, et al. GEANT4—a simulation toolkit. Nuclear Instruments and Methods in Physics Research Section A: Accelerators, Spectrometers, Detectors and Associated Equipment. 2003; **506**:250–303. DOI:10.1016/S0168-9002(03)01368-8

[83] CERN, 2014. Geant4. Available from: http://geant4.web.cern.ch/geant4/support/download.shtml [Accessed: 2017-01-26]

[84] Knoll G. Radiation Detection and Measurement. 3rd ed. New York: Willey; 2000

[85] WaveMetrics. 2013. IGORPro. Available from: http://www.wavemetrics.com/products/igorpro/igorpro.htm [Accessed: 2017-01-26]

[86] Kavetskiy A, Yakubova G, Torbert H, Prior S. Continuous versus pulse neutron induced gamma spectroscopy for soil carbon analysis. Applied Radiation and Isotopes. 2015; **96**:139–147. DOI:10.1016/j.apradiso.2014.10.024

[87] Doron O. Simulation of an INS soil analysis system [thesis]. Austin: The University of Texas at Austin; 2007. Available from: http://www.library.utexas.edu/etd/d/2007/dorono09518/dorono09518.pdf [Accessed: 2017-01-26]

Chapter 8

Effects of Gamma Radiation on Essential Oils: A Review

Clináscia Rodrigues Rocha Araújo,
Geone Maia Corrêa, Viviane Gomes da Costa Abreu,
Thiago de Melo Silva, Aura María Blandón Osorio,
Patrícia Machado de Oliveira and
Antônio Flávio de Carvalho Alcântara

Additional information is available at the end of the chapter

http://dx.doi.org/10.5772/intechopen.68764

Abstract

γ-Radiation provides an effective alternative method to reduce or eliminate microbial contamination of medicinal herbs and other plant materials. However, a search in the literature is important to describe the effects of γ-radiation on the content and integrity of secondary metabolites from plants. The present work provides a review of the effects of γ-radiation on extraction yields and chemical composition of essential oils isolated from roots, rhizome and cortex, leaves, fruits, seeds, flowers, and whole plant. In addition, this review describes the effects of γ-radiation on terpenes. The informations in the present work may assist in research about essential oils and dose of γ-radiation that is able to biologically decontaminate without causing chemical changes in secondary metabolites. These reports in the literature can describe the behavior of many of these metabolites when subjected to various doses of radiation.

Keywords: essential oil, γ-radiation, secondary metabolites, terpenes

1. Introduction

Essential oils (EOs) are plant secondary metabolites, mainly constituted by a mixture of terpenes and terpenoid derivatives (**Figure 1**). Monoterpenes and sesquiterpenes are usually their major constituents [1]. These components are volatile, usually exuding characteristic and pleasant odors. The attraction of insects caused by smell of these essential oils is one of the main factors responsible for the pollination of plants. EOs have many other biological functions, such as protection of plants against diseases caused by fungi and bacteria [2]. Moreover, EOs exhibit a broad spectrum of biological properties, such as mucolytic,

Figure 1. Secondary metabolites present in the different parts of the plant.

expectorant (for example, menthol), and antineoplastic actions [1], stimulating blood flow (for example, EO of mountain pine or common juniper), treatment of gastrointestinal diseases (for example, essential oils of anise, caraway, or fennel), and used in aromatherapy [2]. EOs are also used for production of perfumes and other cosmetic products and are added to foods to improve the flavor [2].

Chemical composition of EOs from plants usually provides important information to its taxonomic identification [3]. However, some environmental factors, such as temperature variation, photoperiod, and light intensity, can influence the biosynthesis of volatile compounds and as a consequence, change its quality and chemical composition [4]. Hydrodistillation is a largely used method in laboratory to obtain EOs from vegetal species. On the other hand, the most usual and popular method to obtain these chemical constituents is a simple decoction. Pressure, temperature, time, dynamic of extraction, and solvent volume are experimental parameters that influence the extraction efficiency and quality of EOs isolated from natural products [5].

Although EOs obtained from plant materials have important medicinal and industrial applications, herbs rich in EOs are often contaminated with microorganisms. Fungal and bacterial contamination is generally caused by the presence of these microorganisms in soil, water, air, or dust, during harvesting, storage, or processing of herbs [6]. The most usual methods for the microbial decontamination of vegetal material are based on applications of ethylene oxide or methyl bromide. However, both methods promote formation of toxic products and have been banned in many countries, such as Japan and those of the European Union [7].

On the other hand, γ-radiation provides an effective alternative method for reducing or eliminating microbial contamination of medicinal herbs and other vegetal products. This type of radiation of high energy usually passes through skin and soft tissue. A small percentage of γ-radiation is absorbed by cells. Once absorbed by a biological material, γ-radiation can provoke direct and indirect effects at molecular level. Direct effects are responsible for DNA double-strand breaks (DSBs), highly toxic lesions that can cause genetic instability and cell death [8].

Indirect effects are more frequent than direct ones. These effects are caused by the interaction of ionizing radiation with water molecules, generating free radicals. In turn, these free radicals are highly reactive with different cell components, such as DNA, enzymes, and secondary metabolites (including EOs) [7]. Literature describes the effects of γ-radiation on the chemical composition of many vegetal species. Chromatographic analyses of EOs isolated from plant indicated that γ-radiation changes their extraction total yields and chemical compositions. Both the parameters (that is, yields and chemical composition) are mainly influenced by vegetal species, radiation dose, and chemical constituents of the plant material [3]. The present review describes the effects of γ-radiation on the chemical composition of EOs isolated from different plant parts: roots, rhizome and cortex, leaves, fruits, seeds, flowers, and whole plant. One entire section is dedicated for description of the effects of γ-radiation on terpenes.

2. Essential oils from roots

Root samples of *Angelica gigas* Nakai (Danggui) purchased from a local market in Korea were submitted to microbial decontamination using γ-radiation at doses of 1.0, 3.0, 5.0, 10.0, and 20.0 kGy. The extraction yields of its volatile oils were 0.314, 0.313, 0.310, 0.312, and 0.290%, respectively. These values are similar to extraction yield obtained for a nonirradiated sample (yield = 0.313%). The profile of volatile components did not change significantly with radiation [3]. The total content of hydrocarbons and monoterpenes in an irradiated sample (average values considering all irradiated materials = 57.2 and 41.8%, respectively) decreased in relation to a nonirradiated sample (60.00 and 44.14%, respectively). The total content of alcohols, sesquiterpenes, and oxygenated sesquiterpenes in irradiated samples (average values considering all irradiated materials = 30.3, 4.7, and 18.5%, respectively) was greater than the corresponding nonirradiated sample (23.35, 4.03, and 13.21%, respectively). The total content of monoterpenes α-limonene, *p*-cymene, and camphene in the irradiated samples (average values considering all irradiated materials = 5.2, 1.2, and 4.7%, respectively) are higher in relation to corresponding nonirradiated samples (4.29, 1.07, and 4.10%, respectively). On the other hand, the content of monoterpenes 2,4,6-trimethylheptane and α-pinene in the irradiated samples (average values considering all irradiated materials = 7.9 and 26.9%, respectively) was lower in relation to corresponding nonirradiated samples (13.39 and 30.89%, respectively). The content of oxygenated monoterpenes verbenol, verbenone, α-eudesmol, β-eudesmol, (*E*)-*p*-2-menthen-1-ol, and pinocarveol in irradiated samples (average values considering all irradiated materials = 2.8, 1.8, 2.5, 7.6, 1.5, and 1.4%, respectively) are higher in relation to corresponding nonirradiated samples (2.15, 1.44, 1.90, 5.01, 1.17, and 1.22%, respectively). Irradiated samples showed higher content of the sesquiterpene α-muurolene (average values considering all irradiated materials = 1.9%) than the corresponding nonirradiated samples (1.52%) [3].

Dried root samples of *Glycyrrhiza glabra* Radix (Licorice) collected in Korea were submitted to γ-radiation at doses of 5.0, 10.0, 25.0, and 50.0 kGy. Irradiated samples exhibited higher total content of aldehydes and hydrocarbons (average values considering all irradiated materials = 24.8 and 17.3%, respectively) than the corresponding nonirradiated samples (17.11 and 15.88%, respectively). Irradiated samples exhibited lower total content of alcohols, ketones, and ethers (average values considering all irradiated materials = 12.8, 2.7 and 5.3%, respectively) than the corresponding nonirradiated samples (14.68, 16.08, and 14.4%, respectively). The total sesquiterpene content in nonirradiated and irradiated samples was near 10.0%. The total monoterpene content was near 4.0% in irradiated and nonirradiated samples, except in the sample irradiated at 50.0 kGy that showed monoterpene content near 8.0% [9].

The total content of aldehydes and hydrocarbons from other materials of the same species (roots of *G. glabra*) collected in Korea was higher in irradiated samples at doses of 5.0, 10.0, 25.0, and 50.0 kGy (average values considering all irradiated materials = 42.4 and 26.2%, respectively) than the corresponding nonirradiated sample (28.69 and 11.31%, respectively). The irradiated samples exhibited lower total content of alcohols and ketones (average values considering all irradiated materials = 2.8 and 2.5%, respectively) than the corresponding nonirradiated sample (9.66 and 14.08%, respectively). The total content of monoterpenes in irradiated samples was higher than for a nonirradiated sample. The total content of sesquiterpenes in irradiated samples was lower than for a nonirradiated sample. The identified volatile components were exactly the same in irradiated and nonirradiated samples of *G. glabra* [9].

Dried and powdered roots of *G. glabra* collected in Syria were submitted to γ-radiation at doses of 5.0, 10.0, 15.0, and 20.0 kGy. Higher contents of glycyrrhetinic acid were observed for irradiated samples (average values considering all irradiated materials = 6.1%) than the corresponding nonirradiated sample (4.37%). The content of glycyrrhetinic acid in both irradiated and nonirradiated samples decreased after 12 months of storage [10].

Roots of *Glycyrrhiza uralensis* Fischer collected in Gwangju (South Korea) were irradiated at doses of 1.0, 3.0, 5.0, 10.0, and 20.0 kGy. The content of the major constituents 2-ethoxy-1-propanol, hexanal, hexanol, *p*-cymen-8-ol, and γ-nonalactone increased in the irradiated samples (average values considering all irradiated materials = 27.1, 6.6, 5.1, 2.4, and 2.6%, respectively) in relation to the nonirradiated sample (22.82, 5.69, 4.78, 2.39, and 2.50%, respectively). However, the content of the constituents ethyl acetate, 4-terpineol, and tetradecanol decreased for irradiated samples (average values considering all irradiated materials = 7.3, 5.8, and 1.7%, respectively) in relation to a nonirradiated sample (7.47, 7.58, and 2.06%, respectively). Benzaldehyde was only detected at a dose of 1 kGy. On the other hand, the compounds 3,5-dimethyl octane and phenethyl alcohol were detected only at a dose of 20 kGy [11].

Samples of *Paeonia albiflora* Pallas var. *trichocarpa* Bunge were submitted to γ-radiation at doses of 1.0, 3.0, 5.0, and 10.0 kGy. Maximum yield of EO was obtained for an irradiated sample at 5 kGy (29.91%). Nonirradiated and irradiated samples at 1.0, 3.0, and 10.0 kGy showed yields of 28.14, 25.89, 26.67, and 25.24%, respectively [1]. Gas chromatography (GC) analysis of volatile compounds obtained from irradiated and nonirradiated samples was similar. A total of 54 compounds was identified in the nonirradiated and irradiated samples at 1.0 kGy. Irradiated samples at 3.0, 5.0, and 10.0 kGy exhibited 55 volatile compounds. This new peak on the GC chromatogram of irradiated samples from 3.0 to 10.0 kGy was attributed to 1,3-bis(1,1-dimethylethyl)-benzene. The highest total contents of alcohols and aldehydes were verified for the irradiated samples (average values considering all irradiated materials = 37.7 and 22.1%, respectively) in relation to the nonirradiated sample (34.26 and 21.44%, respectively). The total contents of acids, esters, furans, ketones, hydrocarbons, and terpenoids were not different among irradiated samples and nonirradiated samples [1].

3. Essential oils from rhizomes and cortices

Ethanol extract of rhizomes of *Cassumunar ginger* purchased in Bangkok was submitted to γ-radiation at doses of 10.0 and 25.0 kGy. EO obtained from nonirradiated and irradiated samples did not show significant changes in the content of sabinene, α-terpinene, γ-terpinene, cymene, α-terpinolene, terpinen-4-ol, α-terpineol, β-sesquiphellandrene, ar-turmerone, ar-curcumene, α-turmerone, β-turmerone, 4-(3′,4′-dimethoxyphenyl)but-3-ene, 4-(3′,4′-dimethoxyphenyl)but-1,-3-ene, (*E*)-1-(3,4-dimethoxylphenyl)butadiene, 4-(2′,4′,5′-trimethoxyphenyl) but-3-ene, and 4-(2′,4′,5′-trimethoxyphenyl)but-1,3-ene [12].

Rhizome samples of *Coptis chinensis* purchased from a local market in Korea were submitted to γ-radiation at doses of 5.0, 10.0, 25.0, and 50.0 kGy. Higher total content of aldehydes and hydrocarbons was verified for irradiated samples (average values considering all irradiated materials = 11.4 and 42.7%, respectively) in relation to nonirradiated samples (7.56 and 35.98%, respectively). Higher total content of sesquiterpene hydrocarbons was verified for irradiated samples at 5.0, 10.0, and 25.0 kGy (average values considering irradiated materials = 47.5, 47.5, and 42.5%, respectively) in relation to nonirradiated sample and irradiated sample at 50.0 kGy (35.0% for both the samples) [9].

Another rhizome sample of *C. chinensis* was purchased from a local market in Korea and irradiated at doses of 5.0, 10.0, 25.0, and 50.0 kGy. The total content of aldehydes in the irradiated samples (average values considering all irradiated materials = 23.1%) was similar to the nonirradiated samples (23.33%). Higher content of hydrocarbons was verified for the irradiated samples (average values considering all irradiated materials = 16.8%) in relation to the nonirradiated samples (11.21%). The total content of sesquiterpene hydrocarbons was higher in the irradiated sample at 25.0 and 50.0 kGy (20.0% for both the samples) in relation to irradiated samples at 5.0 and 10.0 kGy (15.0% for both the samples) and nonirradiated sample (12.5%) [9].

Rhizome samples of *Curcuma longa* (turmeric) purchased from a local market in Kerala (India) were submitted to γ-radiation at doses of 1.0, 3.0, and 5.0 kGy. Similar extract yields of volatile oils were obtained for irradiated (1.54, 1.70, and 1.43%, respectively) and nonirradiated samples (1.52%). A total of 23 constituents was identified in the nonirradiated and irradiated samples. Significant changes in the concentration of their constituents were not observed for irradiated and nonirradiated samples [13].

Another rhizome sample of *C. longa* that was also purchased from the local market in India was submitted to γ-radiation at 10.0 kGy. The overall yield of volatile oil did not change for the nonirradiated sample (1.71%) and irradiated sample (1.72%). The gas chromatography/mass spectrometry (GC/MS) chromatograms did not indicate significant changes in the concentration of its major constituents: α-phellandrene, *p*-cymene, 1:8-cineol, β-caryophyllene, ar-curcumene, mixture of zingiberene and β-sesquiphellandrene, nerolidol, mixture of ar-turmerone and turmerone, curlone, and dehydrozingerone [8].

Fresh rhizomes samples of *Zingiber officinale* var. *Bangalore* (ginger) purchased from a local market in India were submitted to γ-radiation at a dose of 0.06 kGy. The overall yield of EO was slightly higher for an irradiated sample (0.17%) in relation to a nonirradiated sample (0.14%). The GC/MS chromatograms did not indicate significant changes in the concentration of its major constituents: camphene, β-phellandrene, mixture of linalool and α-terpeniol, neral, geranial, ar-curcumene, nerolidol, mixture of zingiberene and zingiberol, and mixture of β-sesquiphellandrene and β-bisabolene [14].

Cortex bark samples of *Cinnamomum zeylanicum* purchased from a local market in Korea were submitted to γ-radiation at doses of 5.0, 10.0, 25.0, and 50.0 kGy. The content of alcohols and aldehydes (average values considering all irradiated materials = 48.5 and 1.3%, respectively) decreases in relation to a nonirradiated sample (65.02 and 1.61%, respectively). On the other hand, the content of hydrocarbons in the irradiated samples (average values considering all irradiated materials = 29.9%) increased in relation to the nonirradiated sample (15.79%). Higher content of sesquiterpenes was verified for irradiated samples (average values considering all irradiated materials = 30.0%) in relation to nonirradiated samples (17.0%) [9].

Another cortex bark sample of *C. zeylanicum* purchased from a local market in Korea was submitted to γ-radiation at doses of 5.0, 10.0, 25.0, and 50.0 kGy. The total content of alcohols, aldehydes, and hydrocarbons increased in irradiated samples (average values considering all irradiated materials = 1.8, 76.5, and 14.9%, respectively) in relation to the nonirradiated sample (1.66, 79.31, and 11.71%, respectively). Higher total content of sesquiterpenes was verified for irradiated samples at 25.0 and 50.0 kGy (17.5% for both the samples) in relation to nonirradiated sample and irradiated samples at 5.0 and 10.0 kGy (near 15.0% for both the samples) [9].

4. Essential oils from leaves

Leaf samples of *Echinodorus macrophyllus* purchased from the local market in Belo Horizonte (Brazil) were submitted to γ-radiation at doses of 1.0, 3.0, 5.0, 10.0, and 20.0 kGy. The overall yield of EO extracted from irradiated leaves (0.67, 0.46, 0.72, 0.58, and 0.418%, respectively) was higher than the corresponding yield for nonirradiated samples (0.27%) [15]. The total content of acyclic monoterpenes and sesquiterpene derivatives in irradiated samples (average values considering all irradiated materials = 0.3 and 3.1%, respectively) was increased in relation to nonirradiated samples (0.11 and 1.90%, respectively). On the other hand, the total content of the other chemical classes, such as triterpenes, diterpenes, esters, and carotenoid derivatives, in irradiated samples (average values considering all irradiated materials = 0.7, 2.7, 76.1, and 3.3%, respectively) was lightly decreased in relation to nonirradiated sample (0.99, 2.93, 79.68, and 4.26%, respectively). Irradiated samples exhibited higher content of linalool, α-caryophyllene, drimenol, hexahydrofarnesyl acetone, (*E,E*)-farnesyl acetone, ethyl hexadecanoate, methyl (*Z,Z*)-9,12-octadecadienoate, and methyl (*E,E,E*)-11,14,17-eicosatrienoate (average values considering all irradiated materials = 0.3, 0.7, 1.5, 0.9, 1.4, 1.9, 2.3, and 11.3%, respectively) in relation to nonirradiated sample (0.11, 0.22, 0.66, 0.40, 0.93, 1.22, 1.70, and 7.94%, respectively). On the other hand, the content of dihydroedulan, β-caryophyllene, methyl hexadecanoate, ethyl (Z,Z,Z)-9,12,15-octadecatrienoate, squalene (average values considering all irradiated materials = 0.9, 0.5, 40.0, 6.8, and 0.7%, respectively) exhibited lower relative content when the samples were not exposed to γ-radiation (2.9, 0.8, 44.3, 12.7, and 1.0%, respectively). The samples submitted to γ-radiation did not exhibit significant changes in the content of 10-(acetylmethyl)-(+)-3-carene, (*E*)-nerolidol, methyl (*Z,Z,Z*)-9,12,15-octadecatrienoate, (*E*)-phytol, methyl octadecanoate, ethyl (*Z,Z,Z*)-9,12,15-octadecatrienoate, and ethyl octadecanoate [15].

Leaf samples of *Eucalyptus radiata* purchased from a local market of Tilman (Belgium) were submitted to γ-radiation at a dose of 25.0 kGy. The overall yield of volatile oil did not change for nonirradiated (0.84%) and irradiated samples (0.85%). Both the samples did not exhibit significant differences in the content of α-pinene, eucalyptol, β-myrcene, terpinen-4-ol, sabinene, neral, α-terpineol, linalyl acetate, and β-bisabolone, which were the major constituents identified in its EO [16].

Leaf samples of *Mentha piperita* L. (a natural hybrid between *Mentha spicata* L. and *Mentha aquatica* L.) were submitted to γ-irradiation at doses of 10.0 and 25.0 kGy. The overall yield of EO for nonirradiated and irradiated samples was similar (0.9%). Irradiated samples exhibited changes in the content of menthol and menthone (average values considering all irradiated materials = 51.6 and 14.5%, respectively) in relation to the nonirradiated sample (52.4 and 13.8%, respectively). Irradiated samples exhibited similar content of γ-gurjunene, *neo*-menthol, α-terpinene, α-gurjunene, *neo*-isomenthol 1,8-cineole, and isopulegol, in irradiated samples (average values considering all irradiated materials = 1.4, 6.4, 0.4, 8.8, 6.4, 3.7, and 1.0%, respectively) in relation to the nonirradiated sample (1.4, 6.5, 0.5, 8.9, 6.5, 3.9, and 0.8%, respectively) [17].

Leaf samples of *M. piperita* purchased in Marrakesh (Morocco) were submitted to γ-radiation at a dose of 1.0 kGy. Some differences were observed in the composition of the EO for irradiated and nonirradiated samples. Higher contents of carvone and dihydrocarveol in the irradiated sample (35.88 and 6.95%, respectively) were verified in relation to the nonirradiated sample (31.83 and 3.14%, respectively). Some nonidentified constituents (GC retention times at 5.67, 5.83, and 6.73 min) exhibited a slight increase for the irradiated sample. The content of viridiflorol, carvacrol, carvyl acetate, and D-germacrene was only detected on the GC chromatogram of the nonirradiated sample (5.35, 3.28, 1.30, and 1.25%, respectively). The content of 1,8 cineole, dihydrocarvyl acetate, and a-bourbonene was similar for both the samples [18].

Leaf samples of *Ocimum basilicum* purchased from the local market in São Paulo (Brazil) were submitted to γ-radiation at doses of 10.0, 20.0, and 30.0 kGy. Chromatographic analysis indicated no significant differences between nonirradiated and irradiated samples [19].

The other leaf sample of *O. basilicum* purchased from a local market in Copenhagen (Denmark) was irradiated using doses of 3.0, 10.0, and 30.0 kGy. The content of 1,8 cineole, β-caryophyllene, methylchavicol, methyleugenol, and linalool also did not exhibit significant differences between nonirradiated and irradiated samples [20].

A third work employing irradiated leaves of *O. basilicum* purchased from Egypt was performed using doses of 5.0 and 10.0 kGy. The content of α-cubebene, an isomer of cadinene, eremophyllene, myristicine, α-pinene, camphene, myrcenol, α-terpineol, allylphenol, safrole, methyl eugenol, and β-bisabolene did not change after exposition to irradiation. However, a significant increase was observed in the content of linalool, estragole, *p*-cymene, 1,8-cineole, *cis*-linalool oxide, *cis*-methyl cinnamate, thymol, β-phellandrene, β-pinene, β-myrcene, γ-terpinene, *cis*-*p*-2-menthen-1-ol, and neral for irradiated samples (average values considering all irradiated materials = 27.7, 13.4, 0.2, 6.2, 0.2, 1.2, 2.4, 0.3, 0.6, 0.2, 0.1, 0.4, and 0.3%, respectively) in relation to the nonirradiated sample (14.92, 4.54, 0.03, 5.14, 0.08, 0.66, 0.68, 0.18, 0.46, 0.10, 0.01, 0.21, and 0.18%, respectively). On the other hand, a decrease was observed for the content of *trans*-methyl cinnamate, α-bergamotene, an isomer of cubebene, β-farnesene, γ-cadinene, calamenene, α-caryophyllene, β-caryophyllene, camphor, borneol, terpin-1-en-4-ol, eugenol, spathulenol, δ-cadinol, and α-cadinol when the samples were exposed to radiation (average values considering all irradiated materials = 5.4, 5.7, 1.8, 0.5, 2.6, 0.3, 0.6, 0.9, 0.3, 0.3, 0.7, 10.1, 0.7, 0.7, and 5.3%, respectively) in relation to the nonirradiated sample (9.08, 6.88, 2.42, 0.77, 3.77, 2.31, 0.73, 1.24, 1.41, 0.72, 0.88, 12.13, 1.04, 1.28, and 9.89, respectively) [21].

Leaf samples of *Origanum vulgare* collected in Turkey were submitted to γ-radiation at doses of 5.0, 7.5, 10.0, and 30.0 kGy. The majority of the identified volatile constituents was only slightly affected by the radiation. Irradiated sample at 5.0 kGy did not exhibit changes in relation to nonirradiated sample. Irradiated sample at 10.0 kGy exhibited a significant increase of the content of linalool, hotrienol, sabinen hydrate, *p*-methoxypyridine, α-terpinolene, and two linalool oxide derivatives. Irradiated sample at 30.0 kGy exhibited a significant increase of *p*-methoxypyridine, α-terpinolene, and both the linalool oxide derivatives. On the other hand, a decrease of bicyclogermacrene was observed for irradiated samples [22].

Leaves of *O. vulgare L.* collected in Santiago Valley (Chile) were submitted to γ-radiation at 1.0, 2.0, 3.0, 5.0, 10.0, and 15.0 kGy. The overall yield of EOs obtained by steam distillation was not different for irradiated and nonirradiated samples. The content of α-pinene, sabinene, myrcene, *p*-cymene, ocimene, *cis*-β-terpineol, carvacryl methyl ether, linalyl propionate, thymol, carvacrol, spathulenol, and caryophyllene oxide increased when the samples were exposed to radiation. However, the content of β-phellandrene, α-terpinene, γ-terpinene, *trans*-sabinene hydrate, terpinolene, *cis*-*p*-2-menthen-1-ol, δ-4-carene, l-borneol, terpinen-4-ol, and piperitol decreased for irradiated samples. Moreover, the CG chromatogram registers an increase of unidentified peaks at doses higher than 5.0 kGy. These peaks are registered at higher retention times than the identified volatile components [23].

Leaf samples of *Thymus vulgaris* L. purchased in Ankara (Turkey) were submitted to γ-radiation at doses of 7.0, 12.0, and 17.0 kGy. Irradiated and nonirradiated samples did not exhibit significant changes on the content of limonene, 1,8-cineole, linalool, borneol, terpinen-4-ol, α-terpineol, thymol, carvacrol, β-caryophyllene, caryophyllene oxide, α-terpinene, γ-terpinene, and *p*-cimene. The content of cumin aldehyde was increased for the samples irradiated at 17.0 kGy (0.6%) in relation to the nonirradiated sample (0.08%). The content of myrcene was not detected in the sample irradiated at a dose of 17.0 kGy [24].

Leaf samples of *T. vulgaris thymoliferum* purchased in Tilman (Belgium) were submitted to γ-radiation at a dose of 25.0 kGy. Overall yield of the volatile oils obtained from nonirradiated and radiated samples was similar (1.11 and 1.12%, respectively). The content of α-pinene, β-pinene, myrcene, *p*-cymene, γ-terpinene, linalool, terpinen-4-ol, *cis*-geraniol, thymol, carvacrol, β-caryophyllene, and in caryophyllene oxide in irradiated samples was also similar to the corresponding nonirradiated sample [16].

Other leaf samples of *T. vulgaris* purchased in Copenhagen (Denmark) were submitted to γ-radiation at doses of 3.0, 10.0, and 30.0 kGy and after being stored for a period of one month. The content of α-thujene, α-pinene, myrcene, α-terpinene, *p*-cymene, *trans*-sabinene hydrate, linalol, borneol, thymbol, carvacrol, β-caryophyllene, and caryophyllene oxide did not exhibit changes among irradiated and nonirradiated samples. The content of γ-terpinene in irradiated samples (average values considering all irradiated materials = 0.101%) was slightly higher in relation to the nonirradiated sample (0.098%) [25].

Dried samples of *Allium fistulosum* L. (Welsh onion) collected in Gwangju (South Korea) were submitted to γ-radiation at doses of 1.0, 3.0, 5.0, 10.0, and 20.0 kGy. The content of 2-methyl-2-pentenal, methyl propyl trisulfide, propylene sulphide, and 1-propane-1-thiol increased for the irradiated samples (average values considering all irradiated materials = 12.0, 8.5, 6.3, and 4.5%, respectively) in relation to the nonirradiated sample (6.07, 4.67, 5.37, and 3.92%, respectively). However, a significant decrease was observed in the content of dipropyl trisulfide, 1-propanethiol, 3,5-diethyl-1,2,4-trithiolane, (*E*)-propenyl propyl disulfide, and (*Z*)-propenyl propyl disulphide (average values considering all irradiated materials = 7.4, 11.8, 1.3, 4.2 and 3.4%, respectively) in relation to the nonirradiated sample (23.77, 16.33, 7.31, 7.59, and 6.02%, respectively). Moreover, nonanal was not detected in the nonirradiated sample, but it was detected at a dose of 3.0 kGy [26].

5. Essential oils from fruits

Fruit samples of *Carica papaya* were submitted to γ-radiation at doses from 0.05 to 3.0 kGy. The GC chromatogram of the irradiated samples exhibited a new peak that was identified as phenol. The content of phenol in different irradiated samples exhibited a dose-dependent increase with radiation, being linear in the dose range of 0.1–3.0 kGy [27].

Fruit samples of *Citrus paradise* (grape fruits) were collected in Texas (USA), and 3 days after harvest, the samples were submitted to γ-radiation at doses of 0.15 and 0.30 kGy. Pulp of non-irradiated fresh grape fruits exhibited higher content of D-limonene and myrcene (10.00 and 0.27%, respectively) than fruits exposed to radiation at 0.15 kGy (6.00 and 0.17%, respectively). However, irradiated sample at 0.3 kGy did not exhibit significant changes of D-limonene and myrcene in relation to nonirradiated sample [28].

A sample of grape fruit variety (Rio Red) collected in Texas (USA) was submitted to γ-radiation at doses of 0.00, 0.07, 0.20, 0.40, and 0.70 kGy. The content of β-carotene, limonin-β-D-glucopyranoside, and total carotenoids did not exhibit changes between irradiated and non-irradiated samples. A total of 35 days after harvest, fruit samples exposed to radiation at 0.07 kGy exhibited higher content of lycopene (1.53%) than fruits exposed to 0.70 kGy (1.32%) [29].

Peel samples of *Citrus unshiu* purchased from a local market in Korea were submitted to γ-radiation at doses of 5.0, 10.0, 25.0, and 50.0 kGy. The total content of acids and alcohols increased in irradiated samples (average values considering all irradiated materials = 8.4 and 6.7%, respectively) in relation to nonirradiated sample (6.43 and 5.46%, respectively). On the other hand, the total content of aldehydes and hydrocarbons decreased in irradiated samples (average values considering all irradiated materials = 21.6 and 52.1%, respectively) in relation to the nonirradiated sample (26.38 and 54.35%, respectively). Higher total content of sesquiterpene hydrocarbons was verified for irradiated samples at 25.0 and 50.0 kGy (5.0% for both the samples) in relation to nonirradiated and irradiated samples at 10.0 and 5.0 kGy (2.5% for each sample). Higher total content of monoterpene hydrocarbons was verified for nonirradiated and irradiated samples at 5.0 and 10.0 kGy (50.0% for each sample) in relation to irradiated sample at 25.0 and 50.0 kGy (45.0% for both the samples. The content of limonene and α-terpineol decreased for increasing radiation doses [9].

Another peel sample of *C. unshiu* purchased from a local market in Korea was submitted to γ-radiation at doses of 5.0, 10.0, 25.0, and 50.0 kGy. The total content of acids, alcohols, and aldehydes increased for the irradiated samples (average values considering all irradiated materials = 6.3, 6.2, and 14.1, respectively) in relation to nonirradiated sample (5.79, 5.87, and 10.80%, respectively). On the other hand, the total content of hydrocarbons decreased in the irradiated samples (average values considering all irradiated materials = 58.7%) in relation to nonirradiated sample (62.30%). Higher total content of sesquiterpene hydrocarbons was verified for irradiated samples at 10.0 and 25.0 kGy (32.5% for both the samples) in relation to nonirradiated and irradiated samples at 5.0 and 50.0 kGy (28.0% for each sample). Higher total content of monoterpene hydrocarbons was verified in nonirradiated and irradiated samples at 5.0 and 25.0 kGy (35.0% for each sample) in relation to average values considering all the irradiated materials = 25.0% for both the samples). The content of limonene and α-terpineol decreased for increasing radiation doses [9].

Volatile extract of *Maroc late* (Mature oranges) harvested in Morocco was submitted to γ-irradiation at doses of 1.0 and 2.0 kGy. GC analysis did not indicate significant differences between nonirradiated and irradiated samples at 1.0 kGy. Irradiated fruits at 2.0 kGy exhibited lower content of linalool, citral, and methyl anthranilate (0.59, 0.16, and 0.08%, respectively) in relation to the corresponding content of nonirradiated sample (0.80, 0.24, and 0.13%, respectively). On the other hand, the content of D-limonene (94.17%) was higher than the corresponding content for nonirradiated samples (93.70%) [30].

Fruit samples of *Piper guineense* purchased from a local market in Eziama Ikeduru (Nigeria) were exposed to γ-radiation at a dose of 10.0 kGy. The content of *p*-cymenol was slightly increased in the irradiated sample (0.19%) in relation to nonirradiated sample (0.05%). The content of α-pinene, camphene, β-pinene, sabinene, myrcene, limonene, phellandrene, 1-8-cineole, ocimene, γ-terpinene, β-cymene, terpinolene, *cis*-linalool oxide, α-cubebene, δ-elemene, α-ylangene, camphor, linalool, gurjunene, *p*-caryophyllene, γ-elemene, cardene, epi-β-farnesene, (*Z*)-β-farnesene, humulene, (*E*)-β-farnesene, germacrene-D, zingiberene, himachalene, β-selinene, bicyclogermacrene, δ-cadinene, sesquiphellandrene, cadina-1,4-diene, calamenene, *p*-cymene-ol, α-calacorene, caryophyllene oxide, methyl eugenol, nerolidol, elemol, guaiol, γ-cadinol, bisabolol, β-eudesmol did not exhibit change when the sample was exposed to γ-radiation [31].

Fruit samples of *Piper nigrum* purchased from a local market in Ankara (Turkey) were submitted to γ-radiation at doses of 7.0, 12.0, and 17.0 kGy. The overall yield of EO decreased for the irradiated samples at 17.0 kGy in relation to the nonirradiated sample. Higher content of γ-terpinene and thymol (average values considering all irradiated materials = 0.64, and 0.50%, respectively) was verified for irradiated samples in relation to the nonirradiated sample. Higher content of cumin aldehyde was verified for irradiated samples at 12.0 and 17.0 kGy (14.73 and 8.30%, respectively) in relation to the nonirradiated sample (0.85%). Higher content of carvacrol, caryophyllene oxide, and *p*-cymene was verified for irradiated sample at 17 kGy (1.85, 2.15, and 0.67%, respectively) in relation to the nonirradiated sample (0.30, 0.94, and 0.34%, respectively). The smaller content of β-selinene was verified for irradiated samples at 12.0 and 17.0 kGy (5.42 and 5.81%, respectively) in relation to the nonirradiated sample (6.88%). The smaller content of methyleugenol, α-gurjunene, and valencene was verified for irradiated sample at 12.0 kGy (2.59, 0.17, and 5.82%, respectively) in relation to the nonirradiated sample (3.33, 0.37, and 7.33%, respectively). The content of α-pinene, β-pinene, myrcene, α-phellandrene, δ-3-carene, limonene, linalool, δ-elemene, α-cubebene, β-caryophyllene, α-caryophyllene, cadinene, spathulenol, (−)-aristolene, and (+)(*E*)-bicyclosesquiphellandrene did not exhibit changes between irradiated and nonirradiated samples [24].

Dried berry samples of *P. nigrum* purchased from a local market in Mäspoma (Slovak Republic) were submitted to γ-radiation at doses of 5.0, 10.0, and 30.0 kGy. The qualitative compositions of their volatile oils were similar for nonirradiated and irradiated samples. The content of their constituents did not exhibit significant change for nonirradiated and irradiated samples at 5.0 and 10.0 kGy. However, lower content of β-elemene, α-guaiene, α-humulene, and β-farnesene was verified for samples irradiated at 30.0 kGy. On the other hand, higher content of *trans*-sabinene hydrate, 3,4-dimethylstyrene, cyclohexenol, *p*-cymen-8-ol, terpinen-4-ol, α-terpineol, α-terpineol, eucarvone, piperitenone, piperitone, undecanone, and spathulenol was verified for irradiated samples at 30.0 kGy. The most significant change was observed for irradiated sample at 30.0 kGy, which exhibit an increase of caryophyllene oxide and a proportional decrease of β-caryophyllene in relation to nonirradiated sample [32].

Another sample of black pepper purchased from a local market in Cairo (Egypt) was submitted to γ-radiation at doses of 5.0 and 10.0 kGy. GC analysis of the nonirradiated sample indicated 21 constituents, while the corresponding irradiated samples indicated 16 and 15 constituents for irradiated samples at 5.0 and 10.0 kGy, respectively. The content of α-thujene, Me-chavicol, and Me-salicylic (average values considering all irradiated materials = 4.4, 3.0, and 0.6%, respectively) increased for irradiated samples in relation to nonirradiated sample (1.36, 1.57, and 0.10%, respectively). The content of α-pinene, β-pinene, α-phellandrene, mixture of β-phellandrene and limonene, geraniol, cymene, terphenyllin, and β-caryophyllene in irradiated samples (average values considering all irradiated materials = 0.2, 1.0, 0.5, 35.8, 1.9, 0.6, 0.5, and 1.5%, respectively) decreased in relation to nonirradiated samples (1.04, 6.32, 3.42, 39.92, 15.54, 0.72, 0.73, and 3.67%, respectively). Myrcene and α-terpinene were not detected in samples exposed at a dose of 5.0 kGy. However, higher content of myrcene and α-terpinene was verified for irradiated samples at a dose of 10.0 kGy (0.9 and 2.6%, respectively) when compared to nonirradiated sample (0.64 and 2.06%, respectively). Terpinol and anisole were not detected at 10.0 kGy. However, both the constituents exhibit higher content at 5.0 kGy (6.89 and 0.60%, respectively) in relation to nonirradiated sample (6.59 and 0.57%, respectively). Undecanal was not detected at 10.0 kGy, and this constituent exhibited lower concentration at 5.0 kGy (0.19%) in relation to nonirradiated sample (1.01%). The content of eugenol was decreased at 5.0 kGy (1.10%) and increased at 10.0 kGy (2.02%) when compared with the nonirradiated sample (1.23%) [33].

Immature fruit samples of *Poncirus trifoliata* (poncirin) purchased from a local market in Korea were submitted to γ-radiation at doses of 5.0, 10.0, 25.0, and 50.0 kGy. The total content of alcohols, aldehydes, and hydrocarbons decreased in irradiated samples (average values considering all irradiated materials = 2.3, 1.1, and 47.0%, respectively) in relation to nonirradiated sample (3.42, 3.38, and 48.82%, respectively). The higher total content of sesquiterpene hydrocarbons was verified for irradiated samples at 50.0 kGy (40.0%) in relation to nonirradiated sample and irradiated samples at 5.0 and 25.0 kGy (37.5% for both the samples) and in relation to irradiated sample at 10 kGy (35.0%). The higher total content of monoterpene hydrocarbons was verified for the irradiated sample at 10.0 kGy (15.0%) and nonirradiated sample and irradiated sample at 5.0 kGy (10.0%) in relation to irradiated samples at 25.0 and 50.0 kGy (5.0% for both the samples). The content of limonene and α-terpineol decreased for increasing radiation doses [9].

Another immature fruit sample of *P. trifoliate* purchased from a local market in Korea was submitted to γ-radiation at doses of 5.0, 10.0, 25.0, and 50.0 kGy. The total content of alcohols, aldehydes, and hydrocarbons increased in irradiated samples (average values considering all irradiated materials = 2.0, 21.9, and 35.2%, respectively) in relation to nonirradiated sample (1.83, 19.77, and 32.72%, respectively). The higher total content of sesquiterpene hydrocarbons was verified for irradiated samples at 5.0 and 25.0 kGy (32.5% for both the samples) and at 10.0 and 50.0 kGy (30.0% for both the samples) in relation to nonirradiated sample

(27.5%). The total content of monoterpene hydrocarbons was similar between irradiated and nonirradiated samples (5.0% for both the samples). The content of limonene and α-terpineol decreased for increasing radiation doses [9].

6. Essential oils from seeds

Seed samples of *Prunus armeniaca* (apricot kernel) purchased from a local market in Korea were submitted to γ-radiation at doses of 5.0, 10.0, 25.0, and 50.0 kGy. The total content of acids, aldehydes, and hydrocarbons (average values considering all irradiated materials = 2.9, 19.4, and 26.3%, respectively) increased in irradiated samples in relation to nonirradiated sample (2.81, 19.12, and 17.27%, respectively). The total content of sesquiterpene hydrocarbons was higher for samples exposed to 25.0 and 50.0 kGy (37.5 and 27.5%, respectively) in relation to nonirradiated and irradiated samples at 5.0 and 10.0 kGy (15% for each sample) [9].

Other samples of *P. armeniaca* purchased from a local market in Korea were submitted to γ-radiation at doses of 5.0, 10.0, 25.0, and 50.0 kGy. The total content of alcohols and aldehydes decreased in irradiated samples (average values considering all irradiated materials = 2.1 and 20.9%, respectively) in relation to nonirradiated sample (2.29 and 37.46%, respectively). The total content of hydrocarbons increased in irradiated samples (average values considering all irradiated materials = 38.0%) in relation to nonirradiated sample (33.48%). The total content of sesquiterpene hydrocarbons was higher in irradiated samples at 10.0 and 50.0 kGy (35.0% for both the samples) in relation to nonirradiated sample and irradiated samples at 5.0 and 25.0 kGy (32.5% for each sample). The chromatographic profiles were almost identical in nonirradiated samples and in samples irradiated at low doses (5.0 and 10.0 kGy). However, irradiated samples at higher doses (from 25.0 to 50.0 kGy) exhibited different chromatographic profiles in relation to the nonirradiated sample. Three other volatile hydrocarbons were detected on the GC chromatogram of the samples irradiated at 10.0, 25.0, and 50.0 kGy, which were identified as 1,7,10-hexadecatriene, 6,9-heptadecadiene, and 8-heptadecene [9].

Seed samples of *Cuminum cyminum* L. purchased from a local market in Ankara (Turkey) were submitted to γ-radiation at doses of 7.0, 12.0, and 17.0 kGy. The higher content of cumin aldehyde was verified for irradiated samples (average values considering all irradiated materials = 64.1%) in relation to the nonirradiated sample (59.75%). The content of carvacrol increased in the irradiated sample at 17.0 kGy (1.55%) in relation to the nonirradiated sample (0.47%). The smaller content of α-phellandrene and γ-terpinene was verified for irradiated samples (average values considering all irradiated materials = 0.2 and 4.6%, respectively) in relation to the nonirradiated sample (0.40 and 10.31%, respectively). The smaller content of limonene was verified for irradiated samples at 12.0 and 17.0 kGy (average values considering all irradiated materials = 0.13%) in relation to nonirradiated sample (0.46%). The content of β-pinene decreased in irradiated sample at 17.0 kGy (1.02%) in relation to nonirradiated sample (3.63%). The content of *p*-cimene, 1,8-cineole, linalool, α-terpineol, phellandral, safranal, β-gurjunene, β-caryophyllene, β-farnesene, p-cimene, linalool, acoradiene, γ-terpinene, and carotol did not change when the samples were exposed to γ-radiation [24].

Seed samples of *C. cyminum* purchased from Iran were also submitted to γ-radiation at doses of 10.0 and 25.0 kGy. The overall yield of EO from irradiated and nonirradiated seed samples was similar (1.5% for each sample). The content of α-terpinen-7-al, γ-terpinen-7-al, and sabinene decreased in the irradiated samples (average values considering all irradiated materials = 0.1, 18.5, and 0.6%, respectively) in relation to nonirradiated sample (0.21, 21.48, and 0.64%, respectively). The content of α-thujene and β-pinene slightly increased in irradiated samples (average values considering all irradiated materials = 0.3 and 24.7%, respectively) in relation to nonirradiated sample (0.21 and 24.07%, respectively). The content of cumin aldehyde, *p*-cymene, and α-pinene decreased in sample irradiated at 10.0 kGy (17.98, 10.20, and 1.03%, respectively) in relation to the nonirradiated sample (19.03, 10.20, and 1.03%, respectively), and the content of these compounds increased in irradiated sample at 25.0 kGy (20.51, 17.75, and 1.12, respectively). The content of myrcene and γ-terpinene increased in irradiated sample at 10.0 kGy (0.85 and 21.91%, respectively) in relation to the nonirradiated sample (0.73 and 20.09, respectively), and the content of these compounds decreased in irradiated sample at 25.0 kGy (nondetected and 15.46%, respectively) [34].

Seed samples of *Salvia sclarea* L. (clary sage) purchased from Konya (Turkey) were submitted to γ-radiation at doses of 2.5, 4.0, 5.5, and 7.0 kGy. Irradiated samples exhibited a decrease in the content of β-pinene, limonene, α-terpineol, and amyl alcohol (average values considering all irradiated materials = 9.8, 11.1, 5.3, and 2.5%, respectively) in relation to nonirradiated sample (18.81, 15.60, 6.54, and 4.82%, respectively). On the other hand, the content of 1,4-dichlorobenzene, 2-ethyl-1-hexanol, and linalool increased for an irradiated sample at 2.5 kGy (19.33, 15.59, and 9.05%, respectively) in relation to a nonirradiated sample (14.74, 14.34, and 7.98, respectively). The content of 1-hexanol also increased for a sample irradiated at 5.5 kGy (11.22%) in relation to a nonirradiated sample (10.29%). Compounds 2,2,4,6,6-pentamethylheptane and 2,2-dimethylundecane were only detected in nonirradiated samples (4.02 and 2.86%, respectively) [35].

Seed samples of *Monodora myristica* were submitted to γ-radiation at 15.0 kGy. The effects of γ-radiation on the EO were not significant. The most remarkable change was observed for α-thujene and β-cymene. The content of these monoterpenes in the irradiate sample (16.76 and 9.29%, respectively) was higher than to the nonirradiated sample (7.14 and 7.14%, respectively). On the other hand, terpinolene, α-terpineol, α-cubebene, and caryophyllene were only detected in small amounts in the nonirradiated sample [36].

Seed samples of *Linum usitatissimum* (linseed) were submitted to γ-radiation at doses of 2.5, 4.0, 5.5, and 7.0 kGy. Irradiated samples at 2.5 and 4.0 kGy decreased the content of *p*-xylene, limonene, and styrene in relation to the nonirradiated sample. The content of 1-hexanol, *p*-xylene, and limonene increased in the irradiated sample at 5.5 kGy. Compounds *p*-cymene, benzaldehyde, and nonanol were not detected in the irradiated samples [37].

7. Essential oils from flowers

Dried flower samples of *Lavandula angustifolia* purchased from Tilman (Belgium) were exposed to γ-radiation at a dose of 25.0 kGy. Overall yield of EO from irradiated sample was similar to the corresponding yield for the nonirradiated sample (0.44%). Nonirradiated and irradiated samples exhibited similar qualitative composition of *p*-cymene, *trans*-β-ocimene, *cis*-β-ocimene, linalool, lavandulol, hexyl acetate, linalyl acetate, β-farnesene, neryl acetate, β-caryophyllene, borneol, caryophyllene oxide, and geranyle acetate [16].

Flower samples of *Crocus sativus* L. purchased from Srinagar (India) were exposed to γ-radiation at a dose of 5.0 kGy. Overall yield of EO from irradiated sample was similar to the corresponding yield of the nonirradiated sample (0.6%). The content of safranal; 2,6,6-trimethyl-4-hydroxy-1-cyclohexene-1-carboxaldehyde; 2,6,6-trimethyl-1,4-cyclohexadione; 3,5,5-trimethyl-2-hydroxy-1,4-cyclohexadione-2-ene; 2,5-dimethyl-2-isopropenyl-l-cyclohexanone; 2,4,4-trimethyl-3-carboxaldehyde-5-hydroxy-1-cyclohexanone 2,5-diene; and dihydro-beta-ionene obtained from the irradiated sample (19.56, 0.63, 1.78, 1.22, 0.34, 2.8, and 3.43%, respectively) decreased in relation to the nonirradiated sample (32.93, 1.57, 1.81, 1.67, 0.35, 3.4, and 3.71%, respectively). The content of α-isophorone, ketoisophorone, and 2,4-cycloheptadiene-1-one-2,6,6-trimethyl (6.17, 3.48, and 1.92%, respectively) increased in relation to the nonirradiated sample (5.25, 3.17, and 1.85%, respectively) [38].

Flower samples of *Solanum stipulaceum* collected in Montes Claros (state of Minas Gerais, Brazil) were exposed to γ-radiation at doses of 1.0, 2.5, 5.0, 10.0, and 20.0 kGy. The yield of volatile oil obtained from nonirradiated flowers was slightly increased by γ-radiation. The content of α-copaene, β-elemene, β-caryophyllene, α-humulene, and aromadendrene increased with radiation except when the plant material was exposed at a dose of 20 kGy. The content of δ-cadinene, caryophyllene oxide, and alloaromadendrene oxide-(2) increased with γ-radiation. The content of D-germacrene, γ-gurjunene, and 7-epi-α-cadinene decreased with γ-radiation in relation to the nonirradiated sample [39].

8. Essential oils from buds

Bud-fermented samples of *Camellia sinensis* (oolong tea) purchased from a local market in São Paulo (Brazil) were exposed to γ-radiation at doses of 5.0, 10.0, 15.0, and 20.0 kGy. Principal component analysis indicated that volatile constituents from samples irradiated at 15.0 and 20.0 kGy showed a chromatographic profile more similar to nonirradiated sample than samples exposed to other applied doses [40]. The content of 4-acetyltoluene, geranial, and δ-dodecalactone was not observed only at a dose of 5.0 kGy. Compounds 2-acetylpyrrole and capric acid were identified at 5.0 kGy. (*E*,*E*)-2,4-heptadienal was not observed only at 15.0 kGy. β-ciclocitral; 1,1,6-trimethyl-1,2-dihydronaphthalene; and eugenol were not observed only at 20 kGy. Benzaldehyde; durol; 2,4-nonadienal; (*E*,*E*)-2,4-decadienal; isopiperitenone; and benzyl benzoate were only identified at a dose of 10 kGy. Linalool, safranal, α-ocimene, and (*E*)-2-decenal were only identified at a dose of 15.0 kGy. Guaiacol, 4-ethylphenol, 4-isopropylbenzaldehyde, and isopropyl methoxy pyrazine were only identified at 20.0 kGy. The content of safranal did not change at doses higher than 15.0 kGy [40].

Bud-unfermented samples of *C. sinensis* were exposed at doses of 5.0, 10.0, 15.0, and 20.0 kGy. Irradiated samples exhibited 82 compounds unidentified in the nonirradiated sample. Benzaldehyde, phenylmethanol, phenylacetaldehyde, geranial, 4-vinylguaiacol, spathulenol, phytone, farnesyl acetone, and phytol were only observed at a dose of 5.0 kGy. Compounds 1-tetradecanal, *cis*-geraniol, octadecanal, and *cis*-linalool oxide were not observed only at a dose of 5.0 kGy. Hexanoic acid, acetophenone, *p*-toluol, (*E*)-2-nonenal, 4-decalactone, β-damascenona, 2,6-dimethyl naphthalene, β-ionone, (+)-aromadendrene, farnesol, and benzyl benzoate were only identified in the sample irradiated at a dose of 10.0 kGy. Compound 4-vinylguaiacol was not observed only at a dose of 10.0 kGy. (+)-Aromadendrene, 2,6-dimethyl-naphthalene, 4-decalactone, and *p*-toluol were not observed only at a dose of 15.0 kGy. Butanoic acid and *trans*-2-decenal were only identified at a dose of 20.0 kGy. Benzyl benzoate and caryophyllene oxide were not observed only at a dose of 20.0 kGy [41].

9. Essential oils from overall herbs

Samples of *Rosmarinus officinalis* L. (rosemary) obtained from a local market in Ankara (Turkey) were exposed at doses of 7.0, 12.0, and 17.0 kGy. EO yield was not affected when samples were exposed to γ-radiation. The content of butylbenzene decreased in irradiated samples (average values considering all irradiated materials = 0.6%) in relation to the nonirradiated sample (0.78%). The content of α-terpinene decreased for the irradiated samples at doses of 12.0 and 17.0 kGy (average values considering all the irradiated materials = 0.3%) in relation to nonirradiated sample (0.37%), but was not changed at 7.0 kGy. The content of limonene, geraniol, and carvacrol decreased for a sample irradiated at 17.0 kGy (2.47, 1.13, and 0.24%, respectively) in relation to nonirradiated sample (2.98, 1.85, and 0.62%, respectively). The content of limonene, geraniol, and carvacrol was not changed for irradiated samples at other doses of radiation. Cumin aldehyde was only identified in irradiated samples (average values considering all irradiated materials = 0.7%). The content of 1,8-cineole increased in irradiated sample at 17.0 kGy (38.2%) in relation to nonirradiated sample (30.73%), but was not changed at other radiation doses. The content of α-pinene, camphene, β-pinene, myrcene, α-phellandrene, δ-3-carene, p-cymene, γ-terpinene, 3-pinanone, bornelo, terpinen-4-ol, α-terpineol, verbenone, thymol, and β-caryophyllene was not affected when the samples were exposed to γ-radiation [24].

Samples of *R. officinalis, Nasturtium officinale* (watercress), and *Cynara scolymus* (artichoke) were exposed at doses of 10.0, 20.0, and 30.0 kGy. Absorption spectrum of irradiated EO did not exhibit significant differences in relation to nonirradiated EO [19].

Samples of *Trifolium pratense* L. (clove) purchased from a local market in Mumbai (India) were exposed to γ-radiation at a dose of 10.0 kGy. Extraction yield of EO increased for the irradiated sample (18.88%) in relation to nonirradiated sample (15.25%). The content of benzylalcohol, eugenol, β-caryophyllene, humulene, eugenol acetate, and vanillin did not change for the irradiated sample [42].

Samples of *Elettaria cardamomum* (cardamom) purchased from a local market in Mumbai (India) were exposed to a dose of 10.0 kGy. The overall yield of EO was not changed for irradiated and nonirradiated samples (5.80 and 5.78%, respectively). The higher content of

α-pinene, sabinene, myrcene, limonene, and nerolidol was observed in irradiated samples (0.68, 1.78, 1.48, 1.23, and 1.89%, respectively) in relation to nonirradiated sample (0.36, 1.40, 1.20, 0.94, and 1.20%, respectively). The content of α-terpineol in irradiated samples (45.43%) decreased in relation to nonirradiated sample (49.0%)[42].

Samples of *Myristica fragrans* (nutmeg) purchased from a local market in Mumbai (India) were exposed to a dose of 10.0 kGy. EO extraction yield was similar to nonirradiated and radiated samples (3.12 and 3.47%, respectively). The content of α-terpeniol, 1-terpinene-4-ol, and myristicin in irradiated samples (6.64, 17.67, and 31.72%, respectively) increased in relation to nonirradiated sample (1.0, 8.6, and 5.0%, respectively). The content of sabinene, β-pinene, and elemicin decreased in comparison to nonirradiated sample [42].

Aerial parts of *Zataria multiflora* purchased from a local market in Shiraz (Iran) were submitted to γ-radiation at 10.0 and 25.0 kGy. EO extraction yield was similar to irradiated and nonirradiated samples (4.0% for each sample). The content of thymol and carvacrol in a nonirradiated sample (61.8 and 10.5%, respectively) decreased for irradiated samples at 10.0 kGy (49.3 and 6.6%, respectively) and 25.0 kGy (45.3 and 6.3%, respectively). On the other hand, the content of *p*-cymene and γ-terpinene in a nonirradiated sample (7.5 and 4.4%, respectively) increased for irradiated samples at 10.0 kGy (16.2 and 8.6%, respectively) and 25.0 kGy (17.5 and 9.3%, respectively) [43].

Leaves and flowers of *T. vulgaris* from Morocco were exposed to γ-radiation at 10.0, 20.0, and 30.0 kGy. Carvacrol was the main component, exhibiting an increase of concentration dose dependent with γ-radiation (81.29% for a nonirradiated sample and 84.0% for an irradiated sample at 30.0 kGy). The content of α-pinene was not altered for the irradiated samples. The content of α-thujene and γ-terpinene decreased at 30 kGy of radiation (0.55 and 2.49%, respectively) in relation to the nonirradiated sample (0.82 and 2.77%, respectively). On the other hand, the content of *O*-acetylthymol increased at 30.0 kGy (0.41%) in relation to nonirradiated sample (0.28%). The content of δ-terpinene increased at 20.0 kGy (0.86%) in relation to the nonirradiated sample (0.75%). The content of *p*-cymene increased at 10.0 and 20.0 kGy (average values considering all irradiated materials = 4.6%) in relation to the nonirradiated sample (3.9%). The content of β-humulene increased at 20.0 and 30.0 kGy (average values considering all irradiated materials = 3.21%) in relation to the nonirradiated sample (2.07%) [44].

Leaves and flowers of *Mentha pulegium* collected in Boujdour (Morocco) were submitted to γ-radiation at 10.0, 20.0, and 30.0 kGy. The content of 3-octanol, *p*-mentha-3,8-diene, menthone, isomenthol, and piperitenone increased for irradiated samples (average values considering all irradiated materials = 0.3, 0.2, 2.1, 0.7, and 8.1%, respectively) in relation to the nonirradiated sample (0.23, 0.10, 1.37, 0.43, and 6.54%, respectively). On the other hand, the content of limonene and piperitone oxide decreased for irradiated samples (average values considering all irradiated materials = 1.4 and 0.7%, respectively) in relation to nonirradiated samples (1.59 and 1.82%, respectively). The content of α-pinene increased at 10.0 kGy (0.42%) in relation to nonirradiated sample (0.37%). The content of β-humulene increased at 10.0 kGy (0.37%) and decreased at 20.0 and 30.0 kGy (average values considering all irradiated materials = 0.13%) in relation to nonirradiated sample (0.25%) [44].

10. Terpenes from essential oils

Geraniol and nerol purchased from Fluka (code. 72170) were solubilized with methanol (100 mg/mL) and were exposed to γ-radiation at a dose of 5.1 Gy/min. Nerol was stable until 96 h of exposition to radiation. On the other hand, irradiated geraniol exhibited changes on the GC/MS profile in relation to the nonirradiated geraniol. Geraniol was isomerized in nerol and linalool. The content of linalool increased with radiation time, from 0.0 (at 24 h) to 8.4% (at 120 h). The content of nerol also increased with radiation time, from 0.0 (at 24 h) to 57.0% (at 120 h) [45].

Citronellol purchased from Aromáticos Gama (Mexico) was exposed to 1.45, 6.07, and 10.02 kGy. The content of citronellol decreased in irradiated samples (95.09, 94.77, and 86.95%, respectively) in relation to nonirradiated sample (97.52%). The content of rhodinol (an impurity) decreased in irradiated samples (0.75, 0.76, and 0.85%, respectively) in relation to nonirradiated sample (2.34%). On the other hand, the content of dihydrocitronellol (another impurity) increased in irradiated samples (1.61, 1.75, and 1.75%, respectively) in relation to nonirradiated sample (0.14%). Moreover, the content of citronellal and hydroxycitronellal increased in irradiated samples (average values considering all irradiated materials = 1.3 and 3.4%, respectively) in relation to nonirradiated sample (0.50 and 3.36%, respectively) [46].

Pure standards of α-pinene, phellandrene, *p*-cymene, eucalyptol, limonene, linalool, lavandulol, terpin-4-ol, and linalyl acetate purchased from Fluka were exposed to 25.0 kGy. The content of these compounds did not change when the samples were exposed to radiation [16].

Pure standards of (+)-camphor, 1,8-cineol, *trans*-cinnamaldehyde, eugenol, ethyl hexanoate, α-ionone, (*S*)-(−)-limonene, 2-phenethyl alcohol, and α-terpineol purchased from Aldrich (code. T3407), (±)-linalool purchased from Fluka (code. 51782), and vanillin purchased from Merck (code. 121-33-5) were exposed to 10.0 and 50.0 kGy. The contents of (±)-linalool, α-terpineol, and α-ionone in an irradiated sample (average values considering all irradiated materials = 0.9, 0.9, and 1.3%, respectively) slightly decreased in relation to nonirradiated sample (1.00, 0.97, and 1.40%, respectively). The content of (S)-(−)-limonene in an irradiated sample at 50.0 kGy (1.24%) decreased in relation to nonirradiated sample (1.30%). No significant differences were verified for the contents of (+)-camphor, 1,8-cineol, *trans*-cinnamaldehyde, eugenol, ethyl hexanoate, and (*S*)-(−)-limonene when exposed to radiation at 10 kGy. Moreover, no significant differences were also verified for the contents of 2-phenethyl alcohol and vanillin for irradiated and nonirradiated samples [47].

11. Discussion

Internal and external factors can determine yield and composition of EOs from plant materials. Internal factors are usually genetic, physiological, and evolutionary (stage of maturity). On the other hand, external factors are usually seasonality, circadian rhythm, temperature, water availability, nutrient availability, air pollution, altitude, mechanical stimuli, attack pathogens, or extraction conditions [13].

The effects of gamma radiation on EOs constituents also depend on the different factors, such as radiation dose, dose rate, vegetal species, temperature, and sample state. Exposition to γ-radiation may increase or decrease the extraction yield of EO and the content of their constituents [35].

The higher yield of EO extraction verified for the irradiated samples has been usually attributed to radiation-induced disruption of the cell wall structure, providing a higher extractability of oil from the plant tissues. Moreover, changes on EO extraction yield can be due to a recombination of the radiolytic products with time. Specific effects can be observed on a secondary metabolite in different essential oils even though submitted to the same radiation conditions. The content of a constituent upon radiation is presumably due to its radiation sensitivity at different doses [14].

Detection and identification of radiolytic products are very important in the study of the effect of radiation in plant materials because changes in the chemical structure of some compounds may lead to formation of toxic radical species. However, the identification of these structures on essential oils is not easy. Radiolytic products are in trace amounts and usually undetected by GC/MS or coelute with other oil constituents. Studies of the effect of γ-radiation on pure compounds are useful to understand these degradation processes, but the response of these compounds could be different when they are a part of a vegetal material.

Volatile compounds, such as terpenes and terpene derivatives, are usually majoritary components in EOs from plant material and contain different chemical functional groups. A volatile compound in different EOs is under specific reactional environments and the conditions of each EO submitted to different ways of isomerization, oxidation, and hydroxylation when exposed to γ-radiation provide new compounds [24].

The effects of radiation on the volatile compounds are different when a constituent is contained in different vegetal specie. For example, the monoterpene linalool showed a great sensitivity to γ-radiation in leaf EO from *O. basilicum* [20]. However, this compound was radiation-resistant in leaf EO from *T. vulgaris* [16]. In the same context, the content of carvacrol significantly decreased in aerial part EO from *Z. multiflora* [43] and increased at doses up to 5.0 kGy in the leaf EO from *O. vulgare*, whereas it was not affected by γ-radiation in leaf EO from *T. vulgaris* [16].

Moreover, the content of α-pinene significantly decreased in the EO from *A. gigas* Nakai [3]. However, the content of this volatile compound increased after exposition to γ-radiation in the aerial part EO from *Z. multiflora* [16]. Another example is the menthone, which increased according to the γ-radiation dose increase in the EO from *M. piperita* [17], but its content increased after γ-radiation at 10.0 kGy in the EO from *M. pulegium* and decreased at 20.0 and 30.0 kGy [44].

In spite of the exposition to radiation on secondary metabolites studied a long time ago, new studies are necessary to better understand its effect on cell structure and chemical structure of the constituents of the EOs from vegetal material. In addition, it seems to be interesting to study whether a modification of the structure of some pure compounds (even in trace) could lead to the formation of toxic, long-lived radicals [16].

Acknowledgements

The authors are grateful to Brazilian agencies CNPq, CAPES, and FAPEMIG for the financial support.

Author details

Clináscia Rodrigues Rocha Araújo[1], Geone Maia Corrêa[2], Viviane Gomes da Costa Abreu[1], Thiago de Melo Silva[1], Aura María Blandón Osorio[1*], Patrícia Machado de Oliveira[3] and Antônio Flávio de Carvalho Alcântara[1]

*Address all correspondence to: aurambo@ufmg.br

1 Departamento de Química, ICEx, Universidade Federal de Minas Gerais - UFMG, Belo Horizonte, Brazil

2 Instituto de Ciências Exatas e Tecnologia, ICET, Universidade Federal do Amazonas - UFAM, Itacoatiara, Brazil

3 Departamento de Química, Universidade Federal dos Vales do Jequitinhonha e Mucuri – UFVJM, Diamantina, Brazil

References

[1] Shim SL, Hwang IM, Ryu KY, Jung MS, Seo HY, Kim HY, Song HP, Kim JH, Lee JW, Byun MW, Kwon JH, Kim KS. Effect of γ-irradiation on the volatile compounds of medicinal herb, Paeoniae Radix. Radiation Physics and Chemistry. 2009;**78**:665-669

[2] Müller M, Buchbauer G. Essential oil components as pheromones. A review. Flavour and Fragrance Journal. 2011;**26**:357-377

[3] Seo HY, Kim JH, Song HP, Kim DH, Byun MW, Kwon JH, Kim KS. Effects of gamma-irradiation on the yields of volatile extracts of *Angelica gigas* Nakai. Radiation Physics and Chemistry. 2007;**76**:1869-1874

[4] Fatemi F, Dadkhah A, Rezaei MB, Dini S. Effect of γ-irradiation on the chemical composition and antioxidant properties of cumin extracts. Journal of Food Biochemistry. 2013;**37**:432-439

[5] Fatemi F, Allameh A, Khalafi H, Rajaee R, Davoodian N, Rezaei MB. Biochemical properties of γ-irradiated caraway essential oils. Journal of Food Biochemistry. 2011;**35**:650-662

[6] Soriani RR, Satomi LC, Pinto TJA. Effects of ionizing radiation in ginkgo and guarana. Radiation Physics and Chemistry. 2005;**73**:239-242

[7] Chatterjee S, Variyar PS, Gholap AS, Padwal-Desai SR, Bongirwar DR. Effect of γ-irradiation on the volatile oil constituents of turmeric (*Curcuma longa*). Food Research International. 2000;**33**:103-106

[8] Fanaro GB, Hassimotto NMA, Bastos DHM, Villavicencio ALCH. Effects of γ-radiation on microbial load and antioxidant proprieties in Black tea irradiated with different water activities. Radiation Physics and Chemistry. 2014;**97**:217-222

[9] Kim MJ, Ki HA, Kim WY, Pal S, Kim BK, Kang WS, Song JM. Development of radiation indicators to distinguish between irradiated and non-irradiated herbal medicines using HPLC and GC-MS. Analytical and Bioanalytical Chemistry. 2010;**398**:943-953

[10] Al-Bachir M, Al-Adawi MA, Al-Kaid A. Effect of gamma-irradiation on microbiological, chemical, and sensory characteristics of licorice root product. Radiation Physics and Chemistry. 2004;**69**:333-338

[11] Gyawali R, Seo H-Y, Shim S-L, Ryu K-Y, Kim W, You SG, Kim K-S. Effect of γ-irradiation on the volatile compounds of licorice (*Glycyrrhiza uralensis* Fischer). European Food Research and Technology. 2008;**226**:577-582

[12] Thongphasuk P, Thongphasuk J, Bavovada R, Chamulitrat W. Effects of gamma-irradiation on active components, free radicals and toxicity of *Cassumunar ginger* rhizomes. International Journal of Pharmacy and Pharmaceutical Sciences. 2014;**6**:432-436

[13] Dhanya R, Mishra BB, Khaleel KM. Effect of gamma-irradiation on curcuminoids and volatile oils of fresh turmeric (*Curcuma longa*). Radiation Physics and Chemistry. 2011;**80**:1247-1249

[14] Variyar PS, Gholap AS, Thomas P. Effect of γ-irradiation on the volatile oil constituents of fresh ginger (*Zingiber officinale*) rhizome. Food Research International. 1997;**30**:41-43

[15] Silva TM, Miranda RRS, Ferraz VP, Pereira MT, Siqueira EP, Alcântara AFC. Changes in the essential oil composition of leaves of *Echinodorus macrophyllus* exposed to γ-radiation. Brazilian Journal of Pharmacognosy. 2013;**23**:600-607

[16] Haddad M, Herent MF, Tilquin B, Quetin-Leclercq JL. Effect of gamma and e-beam radiation on the essential oils of *Thymus vulgaris thymoliferum, Eucalyptus radiata,* and *Lavandula angustifolia*. Journal of Agriculture and Food Chemistry. 2007;**55**:6082-6086

[17] Fatemi F, Dini S, Rezaei MB, Dadkhah A, Dabbagh R, Naij S. The effect of γ-irradiation on the chemical composition and antioxidant activities of peppermint essential oil and extract. Journal of Essential Oil Research. 2014;**26**:97-104

[18] Machhour H, Hadrami IE, Imziln B, Mouhib M, Mahrouz M. Microbial decontamination by low dose gamma-irradiation and its impact on the physico-chemical quality of peppermint (*Mentha piperita*). Radiation Physics and Chemistry. 2011;**80**:604-607

[19] Koseki PM, Villavicencio ALCH, Brito MS, Nahme LC, Sebastião KI, Rela PR, Almeida-Muradian LB, Mancini-Filho J, Freitas PCD. Effects of irradiation in medicinal and eatable herbs. Radiation Physics and Chemistry. 2002;**63**:681-684

[20] Venskutonis R, Poll L, Larsen M. Effect of irradiation and storage on the composition of volatile compounds in Basil (*Ocinum basilicum* L.). Flavour and Fragrance Journal. 1996;**11**:117-121

[21] Antonelli A, Fabbri C, Boselli E. Modifications of dried basil (*Ocinum basilicum*) leaf oil by gamma and microwave irradiation. Food Chemistry. 1998;**63**:485-489

[22] Sádecka J, Polovka M. Multi-experimental study of γ-radiation impact on oregano (*Origanum vulgare* L.). Journal of Food and Nutrition Research. 2008;**47**:85-91

[23] Elizalde JJ, Espinoza M. Effect of ionizing irradiation on *Origanum* leaves (*Origanum vulgare* L.) essential oil composition. Journal of Essential Oil Bearing Plants. 2014;**14**:164-171

[24] Kirkin C, Mitrevski B, Gunes G, Marriott PJ. Combined effects of gamma-irradiation and modified atmosphere packaging on quality of some species. Food Chemistry. 2014;**154**:255-261

[25] Gyawali R, Seo H-Y, Lee H-J, Song H-P, Kim D-H, Byun M-W, Kim K-S. Effect of γ-irradiation on volatile compounds of dried Welsh onion (*Allium fistulosum* L.). Radiation Physics and Chemistry. 2006;**75**:322-328

[26] Venskutonis R, Poll L, Larsen M. Influence of drying and irradiation on the composition of volatile compounds of thyme (*Thymus vulgaris* L.). Flavour and Fragrance Journal. 1996;**11**:123-128

[27] Chatterjee S, Variyar PS, Sharma A. Post-irradiation identification of papaya (*Carica papaya* L.) fruit. Radiation Physics and Chemistry. 2012;**81**:352-353

[28] Vanamala J, Cobb G, Loaiza J, Yoo K, Pike LM, Patil BS. Ionizing radiation and marketing simulation on bioactive compounds and quality of grapefruit (*Citrus paradisi* c.v. Rio Red). Food Chemistry. 2007;**105**:1404-1411

[29] Patil BS, Vanamala J, Hallman G. Irradiation and storage influence on bioactive components and quality of early and late season "Rio Red" grapefruit (*Citrus paradisi* Macf.). Postharvest Biology and Technology. 2004;**34**:53-64

[30] Moussaid M, Caillet S, Nketsia-Tabiri J, Boubekri C, Lacroix M. Effects of irradiation in combination with waxing on the essential oils in orange peel. Journal of the Science of Food and Agriculture. 2004;**84**:1657-1662

[31] Onyenekwe PC, Ogbadu GH, Hashimoto S. The effect of gamma-radiation on the microflora and essential oil of Ashanti pepper (*Piper guineense*) berries. Postharvest Biology and Technology. 1997;**10**:161-167

[32] Sádecká J, Influence of two sterilisation ways. Gamma-irradiation and heat treatment, on the volatiles of black pepper (*Piper nigrum* L.). Czech Journal of Food Sciences. 2010;**28**:44-52

[33] Emam OA, Farag SA, Aziz NH. Comparative effects of gamma and microwave irradiation on the quality of black pepper. Zeitschrift für Lebensmittel-Untersuchung und Forschung. 1995;**201**:557-561

[34] Fatemi F, Dadkhah A, Rezaei MB, Dini S. Effect of γ-irradiation on the chemical composition and antioxidant properties of cumin extracts. Journal of Food Biochemistry. 2013;**37**:432-439

[35] Yalcin H, Ozturk I, Tulukcu E, Sagdic O. Effect of γ-irradiation on bioactivity, fatty acid compositions and volatile compounds of clary sage seed (*Salvia sclarea* L.). Journal of Food Science. 2011;**76**:1056-1061

[36] Zareena AV, Variyar PS, Gholap AS, Bongirwar DR. Chemical investigation of gamma-irradiated Saffron (*Crocus sativus* L.). Journal of Agricultural and Food Chemistry. 2001;**49**:687-691

[37] Onyenekwe PC, Stahla M, Adejo G. Post-irradiation changes of the volatile oil constituents of *Monodora myristica* (Gaertn) Dunal. Natual Product Research. 2012;**26**:2030-2034

[38] Yalcin H, Ozturk I, Hayta M, Sagdic O, Gumus T. Effect of gamma-irradiation on some chemical characteristics and volatile content of linseed. Journal of Medicinal Food. 2011;**14**:1223-1228

[39] Osorio AMB, Silva TM, Duarte LP, Ferraz VP, Pereira MT, Mercadante-Simões MO, Evangelista FCG, Sabino AP, Alcântara AFC. Essential oil from flowers of *Solanum stipulaceum*: Composition, effects of γ-radiation, and antileukemic activity. Journal of Brazilian Chemical Society. 2015;**26**:2233-2240

[40] Fanaro GB, Duarte RC, Santillo AG, Pinto e Silva MEM, Purgatto E, Villavicencio ALCH. Evaluation of γ-radiation onoolong tea odor volatiles. Radiation Physics and Chemistry. 2012;**81**:1152-1156

[41] Fanaro GB, Duarte RC, Araújo MM, Purgatto E, Villavicencio ALCH. Evaluation of γ-radiation on green tea odor volatiles. Radiation Physics and Chemistry. 2011;**80**:85-88

[42] Variyar PS, Bandyopadhyay C, Thomas P. Effect of γ-irradiation on the volatile oil constituents of some Indian spices. Food Research International. 1998;**31**:105-109

[43] Fatemi F, Asri Y, Rasooli I, Alipoor Sh.D, Shaterloo M. Chemical composition and antioxidant properties of γ-irradiated Iranian *Zataria multiflora* extracts. Pharmaceutical Biology. 2012;**50**:232-238

[44] Zantar S, Haouzi R, Chabbi M, Laglaoui A, Mouhib M, Boujnah M, Bakkali M, Zerrouk MH. Effect of gamma irradiation on chemical composition, antimicrobial and antioxidant activities of *Thymus vulgaris* and *Mentha pulegium* essential oils. Radiation Physics and Chemistry. 2015;**115**:6-11

[45] Srivastava P, Wagh RS, Naik DG. γ-Irradiation: A simple route for isomerization of geraniol into nerol and linalool. Radiochemistry. 2010;**52**:561-564

[46] Rodríguez-Linares D, Codorniu-Hernández E, Aguilera-Corrales Y, Rapado-Paneque M, Quert-Alvarez R, Noel N. Structural changes on citronellol induced by gamma-radiation. Journal of Physical Organic Chemistry. 2009;**22**:527-532

[47] Sjövall O, Honkanen E, Kallio H, Latva-Kala K, Sjöberg AM. The effects of gamma-irradiation on some pure aroma compounds of spices. Zeitschrift für Lebensmittel-Untersuchung und Forschung. 1990;**191**:181-183

Chapter 9

Gamma Rays' Effect on Food Allergen Protein

Marcia Nalesso Costa Harder and Valter Arthur

Additional information is available at the end of the chapter

http://dx.doi.org/10.5772/67839

Abstract

Many foods cause different kinds of allergies for so many people due to health problems. Recently, gamma rays have been used to minimize this problem by altering the protein allergen structure. The aim of this study is to represent the use of the gamma rays in allergen food treatment and to show what happened to food structures. It can be concluded that the use of the technique of irradiation by gamma rays may be an efficient solution for allergic foods.

Keywords: gamma radiation, food irradiation

1. Introduction

Many types of food induce different allergies and may lead to severe health problems.

Basically, preventing the allergic disorders requires to stop eating the food promoter of the allergic reaction. The food irradiation by specific ionizing radiation is used to improve the safety and storability and is one of the most extensively studied technologies since the radiation discovery.

There are some attempts to use gamma rays to minimize this problem by changing the protein allergen structure like some research that show this effect in eggs and milk allergenic protein, in shrimps, and in other allergic foods. These researches showed that the gamma radiation could minimize the quantity of the allergen by altering the promoter protein structure; this step may depend on the radiation dose.

The aim of this work is to show how to employ radiobiology for peaceful purposes and to help allergic people to consume more healthy and safe food.

2. Food allergy

Food is essential for life, usually a source of pleasure, and is usually related to the cultural identity of each population. Over a lifetime, a person eats about two or three tons of food. Not surprisingly, therefore, that diet is implicated in a wide variety of symptoms leading to disorders in the lives of individuals who believe they have food allergies [1]. Also according to the same authors, the first description of food allergy was made by Hippocrates more than 2000 years ago, but only from the 1980s extensive studies have begun to shed light on the different aspects of food allergy. Meanwhile, numerous publications led us to attribute any symptom in human to a certain allergic reaction to a particular food.

Etymologically, the word ALLERGY derives from the word ALLOS, which in Greek means altered state/other, and ERGON, which means energy/reaction, and was first introduced in 1905 by Austrian pediatrician Clemens von Piquet (1874–1924) to designate a clinical manifestation [2]. Moreira [3] defined allergy as a chronic disease characterized by an increase in lymphocyte capacity B to synthesize immunoglobulin isotype IgE antigens that enter the body by inhalation, ingestion, or penetration of the skin leading to an immune hyperactivity, allergic inflammation, and with 20% loss of body weight, when tested in animals process.

Already Pereira et al. [4] defined food allergy as an adverse reaction to a food antigen-mediated immune mechanisms fundamentally. They also reported that it is a nutritional problem, which showed a considerable growth in the last few decades probably due to greater exposure of the population to a larger number of available food allergens, it has become a health problem worldwide because it is associated with a negative impact on quality of life.

In general, allergy is a hypersensitivity reaction initiated by immunological mechanisms. Antibodies or cells may mediate the allergy. In most cases, the antibody responsible for allergic reaction belongs to the IgE isotype and those individuals are referred to as "suffering from an IgE-mediated allergy." Not all IgE-associated allergic reactions occur in atopic individuals. In non IgE mediated allergy, the antibody can belong to the IgG isotype, i.e., serum sickness previously referred to as reaction type III [5].

According to the same author, allergens are antigens, which cause allergy. Many allergens to react with IgE and IgG are proteins, often with carbohydrate chains that under certain circumstances they have been referred to as allergens themselves. Rarely, chemicals of low molecular weight such as isocyanates and anhydrides, which act as haptens are referred to as allergens.

Adverse reactions to food (ARF) is the name used for any abnormal reaction to food or food additives, regardless of their cause. These can be classified into intolerance and food hypersensitivity. Food intolerance is the term for an abnormal physiological response to food or food additives, not from immune nature. These reactions may include: metabolic abnormalities responses to pharmacological substances contained in foods, toxic reactions, among others. Food hypersensitivity and food allergy (FA) is the term used for the ARF, involving immunological mechanisms resulting in great variability of clinical manifestations. The mechanism involving immunoglobulin E (IgE) is the most commonly involved, which is characterized by rapid installation and clinical manifestations such as urticaria, bronchoconstriction, and possibly anaphylaxis. When no immune responses measured by IgE are involved, the clinical manifestations are established later (hours or days), making the diagnosis of FA [6].

When the ARF are caused by immunological mechanisms are said to FA, whereas when caused by toxic, pharmacological, metabolic, or idiosyncratic reactions to chemicals are said food intolerances [3, 6]. For this reason everything has mistakenly been considered allergies and intolerances as synonyms, and many of these adverse effects are blamed for promoting allergic processes [4, 7] and what are the main food allergens are nature protein [4].

Food allergy can begin in childhood and usually arises when the family has a history of atopic diseases (such as allergic rhinitis or allergic asthma). The first hint of allergic predisposition may be a rash as eczema (atopic dermatitis). Such rash may or may not be accompanied by gastrointestinal symptoms, such as nausea, vomiting, and diarrhea, and may or may not be caused by a food allergy [8].

According to the same author, children with allergies to certain foods probably will contract other atopic diseases as they grow as allergic asthma and seasonal allergic rhinitis. However, in adults and children over 10 years, it is very unlikely that food is responsible for respiratory symptoms, despite skin tests (skin) being positive. People who are allergic to foods may react violently to eat a minimal amount of the substance in question, can be covered with a rash all over the body, feel the throat ignite until it closes, and have breathing difficulties. A sudden drop in blood pressure can cause dizziness and collapse. This emergency, life-threatening problem is called anaphylaxis. Some people just suffer from anaphylaxis when exercising immediately after eating the food to which they are allergic.

And in recent decades, there has been an increase of allergic problems promoted by food in children and young people [4, 9]. This contributes negatively to the quality of life for the population and becomes a public health problem that affects the whole world [3, 4, 10, 11]. There is no specific treatment for food allergies, instead is necessary to stop eating foods that trigger.

Type I allergy is characterized by immunoglobulin E (IgE), which mediates hypersensitivity afflicting more than one fourth of the world population [1].

About 2.5% of the adult population suffers from some type of allergy and certainly unknown. Between 100 to 125 people die each year in the US because of an allergic reaction to foodborne [4, 12].

Anaphylaxis is the most severe allergic conditions seen by allergists and may quickly lead to death of healthy individuals and according to Lopes [13], ingestion of food allergens is a major trigger of anaphylactic reactions. Among the most allergenic foods for children are cow's milk and egg white, and for adults are crustaceans [13, 14].

3. Gamma radiation use in food

Radiation is a form of energy. It is in the form of atomic particle or electromagnetic energy such as alpha particles, electrons, positrons, protons, neutrons, etc., which can be produced in particle accelerators or reactors and alpha particles; electrons and positrons are also spontaneously emitted from the nuclei of radioactive atoms [15].

The sources of Cobalt 60 (Co^{60}) ($T_{1/2} = 5.263$ years), using Co^{60} source, whose gamma rays have greater penetration power more than that from electron beams, the objectives of irradiating food can be achieved [16, 17]. The gamma rays are applied in large thickness or bulk foods, while the electron beam used in superficial irradiation [18].

According to the Codex Alimentarius [19], the following radiation sources are allowed:

(a) Gamma rays radionucleotides as Cobalt 60 (Co^{60}), and Cesium 137 (Ce^{137});

(b) X-rays generated by machines operating with power up to 5 MV;

(c) Electrons generated by machines operating at 10 MV or below.

Cardoso [20] says that unfortunately the great benefits of nuclear energy are little known. According to the author, every day, new nuclear techniques are developed in various fields of human activity, making it possible to perform tasks impossible to be carried out by conventional means. The research, industry, health, pharmaceutical, and particularly agriculture are the most benefited areas.

According to Nouailhetas [21], the use of ionizing radiation for the benefit of man, related to human health became evident. Throughout history, these effects have been identified and described mainly from situations in which man is found exposed acutely, either by accident or during medical use. Effects that may result from exposure to radiation in natural conditions have been little studied and poorly understood. Recently, studies have been conducted in order to better understand the role of these radiations by the life has been developed and is expected to issue new concepts about the biological effects of ionizing radiation.

When interacting with matter, different types of radiation can produce various effects that can be simply a color sensation, a perceived sensation of heat, and an image obtained by radiography. Radiation is called ionizing when it produces ions, radicals, and free electrons in irradiated matter. Ionization occurs when radiation has energy enough to break chemical bonds or expel electrons of atoms after collisions.

Due to the global needs of the food security and the problems, arising from inadequate storage and inadequate processing, contemporary agribusiness runs in constant development of new methods of preservation of agricultural products. Irradiation is a method of preservation on both levels: raw materials in nature and adjunct industrial processes [22].

The use of ionizing radiation in all lines of research such as agriculture, health, and industry generates much controversy due to mainly several nuclear accidents worldwide. Therefore, radiation for the population already may be a risk and harm to human health. People should know the difference between contaminated material with radioactive elements and materials processed by irradiation. Irradiators used for food processing are well secured. All safety rules and procedures are followed, which reduce the risks of using these devices in food irradiation process serving for disinfection, sterilization, control of insects and increase the shelf life of foods, delay in fruit ripening, inhibit sprouting, etc., without causing substantial damage to food. It is important to convince consumers that irradiated food does not become radioactive and toxic waste and can be consumed immediately after irradiation process [23].

According to the same author, irradiation is an effective technique in food preservation because it reduces natural losses caused by physiological processes such as sprouting, ripening and aging, and eliminating or reducing microorganisms, parasites, and pests without causing any damage to food, making them also safer for the consumer. The process consists of submitting them already packaged or in kind to a controlled amount of radiation for a given time and set goals. Irradiation can prevent the multiplication of microorganisms such as bacteria and fungi that cause deterioration or losses in products not only by altering the molecular structure but also by inhibiting the maturation of some fruits and vegetables by changes in the physiological processes of the plant tissue.

For Landgraf [24], there is no doubt that the main objective of the use of ionizing radiation in food is to make them safe by reducing the population of pathogenic microorganisms and deteriorating to undetectable levels, and thus increasing the life of products and ensuring their safety. This technology can also be used for other purposes such as to inhibit sprouting of onions and potatoes, insect infestation control, quarantine treatment, and change of allergenic foods, among other applications.

The irradiation has the following advantages over treatment with chemicals, heat treatment or a combination of both. It is a continuous and very efficient process that ensures complete disinfection of products. It leaves no residue in fruits and tends to slow the ripening of climacteric fruit without deteriorating storage time [25].

Importing and exporting countries have shown interest in irradiation technology, conducting research for this application due to increased international trade in food and the growing regulatory requirements of consumer markets [26].

The effects of irradiation on food depend on the food type, the kind of radiation, and its dose. The irradiated products can be transported, stored, and consumed immediately after treatment. In 1983, a working group from the United Nations established standards and set the dose limits; they recommended 10 kGy as the upper limit in food irradiation [27, 28].

The absorption of electromagnetic radiation by biological tissues that constitute the food is a function of the electronic excitability of the constituent molecules. In the case of gamma radiation, the electronic excitation produced is sufficient to eject electrons from orbitals, resulting in their molecular ionization. One of the most important free radicals induced by radiation is the formation of the hydroxyl radical (HO—) that is involved in the initiation and propagation reactions [29].

Food irradiation was supported by FAO/WHO, being that the FDA, in 1963, considering the irradiation of food a safe process, as well as an important tool for preventing poisoning and food infections, approving its use for a variety of foods [13, 30, 31].

For all food submitted to food irradiation, the Radura symbol can be used to identify the process (**Figure 1**), which should be placed on irradiated food packages in many countries of the world. The Radura symbol originated from and was copyrighted by an irradiation

Figure 1. The Radura symbol.

food processing facility located in Wageningen, Netherlands in the 1960s. Then, President Jan Leemhorst of the company called Gammaster recommended its use as an international label to be placed on irradiated food as long as manufacturers implemented appropriate quality parameters. The Radura symbol is listed in the Codex Alimentarius Standard on Labelling of Prepackaged Food. The FDA requires that foods that have been irradiated bear the "Radura" logo along with the statement "Treated with radiation" or "Treated by irradiation." [31].

4. Irradiation of allergenic proteins

Irradiation may cause irreversible changes in the molecule by breaking covalent and conformational changes. Therefore, it is an attractive option to cause the breakdown of the possible allergenic structures naturally present in various foods [32–35].

Gutierrez et al. [36] found that irradiating a cheese dish, which was in the maturation of casein proteolysis yields compounds of low molecular weight and those with increased irradiation doses showed decreased percentage of non-protein nitrogen.

Harder [37] introduced the use of gamma radiation to minimize the effects of allergenic ovomucoid laying hen eggs protein and noted the potential of this technology for this purpose. According to Ma et al. [38], the yolk protein is more susceptible to breakage than proteins of white egg. The chemical change of the radiation in the egg and yolk powder, irradiated in the presence of oxygen, induces degradative changes of the lipid components: a lipid hydroperoxides accumulation and the destruction of carotenoids. For the food industry, this is as important as the deterioration of functional proteins such as egg withe powder foam and the emulsifying ability of the yolk.

Jankiewicz et al. [39] studied the stability of the immunochemical allergen in celery roots, including gamma radiation (10 kGy) and concluded that the allergenicity was only slightly reduced at this dose. Similarly, in studies with cashew nuts, Su et al. and Venkatachalam et al. [40, 41] used doses up to 25 kGy where the major allergens showed stability. In contrast, Byun et al. [42] found effects of decreasing allergenicity of potentially allergenic foods as increasing dose of gamma radiation used, aiming beta-lactoglobulin protein in cow's milk, hen egg albumin, and shrimp tropomyosin using doses up to 10 kGy as shown in (**Figure 2**).

Sinanoglu et al. [43] completely inactivated the allergen shrimp tropomyosin by irradiating them at a dose of 4.5 kGy confirming the results of Zhenxing et al. [44] but found that at doses lower than 5 kGy allergenicity increases. Leszczynska et al. [45] used doses up to 12.8 kGy in wheat flour gliadin irradiation and noted small increase in ω-gliadin as the dose of radiation increases.

Although similar amino acid irradiation cannot be directly extrapolated to peptides or proteins, in any event, the contribution is observed that more complex amino acids and peptides provide valuable insights in the study of more complex proteins. A lesser extent of free amino acids and peptides is also present in food and meat (approximately 2%), and therefore, they are treated as separate food components and not merely as "building blocks" of proteins. The soluble fraction of the flesh, the sarcoplasm consists of approximately 25–30% of the total protein content. It consists of enzymes (glycolytic, etc.), myoglobin, amino acids, nucleotides (ATP, ADP, NAD, FAD, etc.), carbohydrates (glucose intermediates), lactic acid, etc.

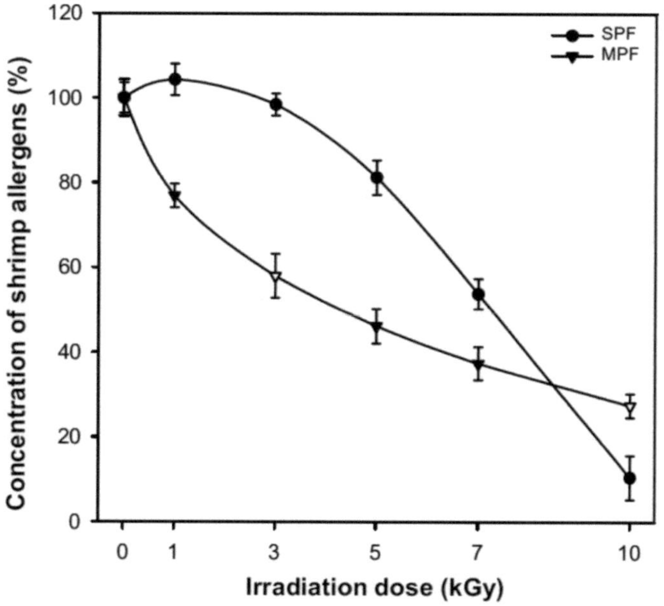

Figure 2. Concentration of allergens in gamma-irradiated shrimp extracts detected by shrimp-hypersensitive patients' IgE. The concentration was measured by CI-ELISA formatted with shrimp tropomyosin as standard allergen. SPF and MPF indicate sarcoplasmic protein fraction and myofibrillar protein fraction, respectively [42].

In view of limited number of detailed studies that have been conducted, only few can cause widespread attempts over protein-OH reaction. The fraction of highest reactivity is located exactly on the aromatic and heterocyclic amino acid in the protein. Therefore, more than 3% of OH radicals are involved in the separation of the C—H bond. Approximate calculations regarding the reactivity of the protein can be applied using constant proportions to amino acid constituents. The distribution of attack, the entire route can depend on the conformation of the protein. Deeply groups as —SS— are expected to be less accessible to the OH radical. The experimental determination of *k* value for OH radical reaction, i.e., alcohol dehydrogenase, catalase, lysozyme ribonuclease, tripisin, trypsinogen, etc., is too high, on the order of 1011 m^{-1} s^{-1}, mainly for its broad size, which increases the frequent meeting and multiplication of reactive sites. The radiation interaction may lead to protein deamidation, breakage of peptides, and disulfide bond, in addition to aromatic residues, heterocyclic amino acids, and metmyoglobin reduction, and the extent and rate of these reactions depend on the conditions and the proteins in this system. These reactions are endless for the variety of radiological products found in irradiated foods. In freezing systems, the OH radical contribution will greatly decrease and only a small fraction of the OH radical inducing products will be present in radappertization food. However, recognition of those products is desirable [46].

In general, there are few researches that have focused on the use of gamma radiation to reduce the allergenicity in food taking into account the protein kind or extrapolating this to the amino acid level. There is a real necessity for additional studies to define the suitable conditions to submit the food for the optimum gamma ray treatment.

Author details

Marcia Nalesso Costa Harder[1*] and Valter Arthur[2]

*Address all correspondence to: mnharder@terra.com.br

1 Technology College of Piracicaba "Dep. Roque Trevisan", Piracicaba, Brasil

2 Nuclear Energy Center in Agriculture – CENA/USP, Piracicaba, Brasil

References

[1] Grumach, A.S., editor. Alergia e imunologia na infância e na adolescência. 1st ed. São Paulo: Ed. Atheneu; 2001. 928 p.

[2] Fernandes, M.E. Alergia alimentar em cães [thesis]. São Paulo: USP; 2005. 104 p.

[3] Moreira, L.F. Estudo dos componentes nutricionais e imunológicos na perda de peso em camundongos com alergia alimentar. [dissertation]. Belo Horizonte: UFMG; 2006. 75 p.

[4] Pereira, A.C. S.; Moura, S.M.; Constant, P.B.L. Alergia alimentar: sistema imunológico e principais alimentos envolvidos. Semina: Ciências Biológicas e da Saúde. 2008;**29**(2): 189–200.

[5] Johansson, S.G.O.; Bieber, T.; Dahl, R.; Friedmann, P.S.; Lanier, B.Q.; Lockey, R.F.; Motala, C.; Ortega, M.J.; Platts-Mills, T.A.E.; Ring, J.; Thien, F.; Van Cauwenberge, P.; Williams, H.C. Revised nomenclature for allergy for global use: Report of the Nomenclature Review Committee of the World Allergy Organization. Journal of Allergy and Clinical Immunology. 2004;**113**(4):832–836.

[6] Parker, S.L.; Krondl, M.; Coleman, P. Foods perceived by adults as causing adverse reaction. Journal of the American Dietetic Association. 1993; **93**(1): 40–46.

[7] De Angelis, R.C., editor. Alergias alimentares: tentando entender por que existem pessoas sensíveis a determinados alimentos. 1st ed. São Paulo: Atheneu; 2006. 124 p.

[8] Manual Merck. Alergia e intolerância alimentar. In: Manual Merck. Doenças do sistema imunitário. [Internet]. Available from: http://www.manualmerck.net/?url=/artigos/%3Fid%3D195%26cn%3D1683 [Accessed: jun 2016].

[9] Larramendi, C.H. Proposal for a classification of food allergy. Alergología e Inmunología Clínica. 2003;18(2):129–146.

[10] Ferreira, C.T.; Seidman, E. Alergia alimentar: atualização prática do ponto de vista gastroenterológico. Jornal de Pediatria. 2007;83(1):7–20.

[11] Lopes, C.; Ravasqueira, A.; Silva, I.; Caiado, J.; Duarte, F.; Didenko, I.; Salgado, M.; Silva, S.P.; Ferrão, A.; Pité, H.; Patrício, L.; Borrego, L.M. Allergy, from diagnosis to treatment. Revista Portuguesa de Imunoalergologia. 2006;14(4):355–364.

[12] Sanz, M.L. Inmunidad del tracto intestinal: procesamiento de antígenos. Alergologia e Inmunologia Clinica. 2001;16(2):58–62.

[13] Lopes, T.G.G. Efeito da radiação gama na reatividade alergênica e nas propriedades físico-químicas e sensoriais de camarão (Litopenaeus vannamei) [thesis]. Piracicaba: USP; 2012. 91 p.

[14] Bernd, L.A.G.; Fleig, F.; Alves, M.B.; Bertozzo, R.; Coelho, M.; Correia, J.; Di Gesu, G.M.S.; Di Gesu, R.; Geller, M.; Mazzolla, J.; Oliveira, C.H.; Peixoto, D.S.A.; Sarinho, E.; Silva, E.G. Anafilaxia no Brasil: Levantamento da ASBAI. Revista Brasileira de Alergia e Imunopatologia. 2010;33(5):190–198.

[15] Okuno, E. Efeitos biológicos das radiações ionizantes: acidente radiológico de Goiânia. Estudos Avançados. 2013;27(77):11–29.

[16] Diehl, J.F., editor. Safety of irradiated food. 1st ed. New York: Marcel Dekker; 1995. 345 p.

[17] Lepki, L.F.S.F. Efeito da radiação ionizante na viscosidade do ovo industrializado [thesis]. IPEN: São Paulo; 1998. 70 p.

[18] Santin, M. Use of irradiation for microbial decontamination of meat: situation and perspectives. Meat Science. 2002;62:277–283.

[19] Codex Alimentarius Commission, editor. Codex general standard for irradiated foods. 15th ed. Rome: Codex Alimentarius; 1984.

[20] Cardoso, E.M., editor. Aplicações da Energia Nuclear. 1st ed. São Paulo: CNEN: Comissão Nacional de Energia Nuclear; 2004.

[21] Nouailhetas, Y., editor. Radiações ionizantes e a vida. 1st ed. São Paulo: CNEN: Comissão Nacional de Energia Nuclear; 2004. 126 p.

[22] Villavicencio, A.L.C.H. Avaliação dos efeitos da radiação ionizante de 60Co em propriedades físicas, químicas e nutricionais dos feijões Phaseolus vulgaris L. e Vignaunguiculata (L.) Walp. [thesis]. São Paulo: USP; 1998. 138 p.

[23] Domarco, R. E. ; Walder, J. M. M. ; Arthur, V. ; Wiendl, F. M. Irradilation of agricultural products to reduce post-havest losses in Brasil. Irradiation of agricultlural products. 1st ed. Vienna: IAEA/FAO, 1992. 32 p.

[24] Landgraf, M. Avanços na Tecnologia de Irradiação de pescados. [Internet]. 2012. Available from: ftp://ftp.sp.gov.br/ftppesca/IIsimcope/palestra_mariza_landgraf.pdf [Accessed: jun 2016]

[25] Moller, J. Projected costs of fall related injury to older persons due to demographic change in Australia: Report to the Commonwealth Department of Health and Ageing. 1st ed. Canberra: New Directions in Health and Safety; 2003. 99 p.

[26] International Atomic Energy Agency. Consumer acceptance and market development of irradiated food in Asia and the Pacific. 1st ed. Vienna: IAEA; 2001. 98 p.

[27] International Atomic Energy Agency. Food irradiation processing. 1st ed. Vienna: IAEA; 1985. 578 p.

[28] International Atomic Energy Agency. Facts about food irradiation. 1st ed. Vienna: IAEA; 1999. 53 p.

[29] Riley, P.A. Free radicals in biology oxidative stress and the effects of ionizing radiation. International Journal of Radiation and Biology. 1994;65(1):27–33.

[30] World Health Organization. Inocuidad y idoneidad nutricional de los alimentos irradiados. 1st ed. Geneva: WHO; 1995. 172 p.

[31] US Food and Drug Administration. Irradiation: A safe measure for safer iceberg lettuce and spinach [Internet]. 2008. Available from: http://www.fda.gov/downloads/ForConsumers/ConsumerUpdates/UCM143389.pdf [Accessed: jun 2016]

[32] US Food and Drug Administration. Foods permitted to be irradiated under FDA regulations (21 CFR 179.26). [Internet]. 2009. Available from: http://www.fda.gov/Food/FoodIngredientsPackaging/IrradiatedFoodPackaging/ucm074734.htm [Accessed: jun 2016]

[33] US Food and Drug Administration. Irradiation and Food Safety: answers to frequently asked questions. [Internet]. 2012. Available from: www.fsis.usda.gov/Fact_Sheets/Irradiation_and_Food_Safety/index.asp [Accessed: jun 2016]

[34] Yang, W.W.; Chung, S.Y.; Ajayi, O.; Krishnamurthy, K.; Konan, K.; Goodrich-Schneider, R. Use of pulsed ultraviolet light to reduce the allergenic potency of soybean extracts. International Journal of Food Engineering. 2010;6(11):79–85.

[35] Kume, T.; Matsuda, T. Changes in structural and antigenic properties of proteins by radiation. Radiation Physics and Chemistry. 1996;46:225–231.

[36] Gutierrez, E.M.R.; Domarco, R.E.; Spoto M.H.F.; Blumer, L.; Matraia, C. Effects of gamma radiation on the physical-chemical and microbiological characteristics in the prato cheese ripening period. Food Science and Technology. 2004;24(4):596–601.

[37] Harder, M.N.C. Efeito do urucum (Bixa orellana) na alteração de características de ovos de galinhas poedeiras. [thesis]. Piracicaba: USP; 2009. 60 p.

[38] Ma, C.Y.; Harwalkar, V.R.; Poste, L.M.; Sahasrabudhe, M.R. Effect of gamma irradiation on the physicochemical and functional properties of frozen liquid egg products. Food Research International. 1993;26:247–254.

[39] Jankiewicz, A.; Baltes, W.; Bögl, K.W.; Dehne, L.I.; Jamin, A.; Hoffmann, A.; Haustein, D.; Vieths, S. Influence of food processing on the immunochemical stability of celery allergens. Journal of the Science of Food and Agriculture. 1997;75:359–370.

[40] Su, M.; Venkatachalam, M.; Teuber, S.S.; Roux, K.H.; Sathe, S.K. Impact of γ-irradiation and termal processing on the antigenicity of almond, cashew nut and walnut proteins. Journal of the Science of Food and Agriculture. 2004; 84 (10):1119–1125.

[41] Venkatachalam, M.; Monaghan, E.K.; Kshiragar, H.H.; Robotham, J.M.; O'Donnell, S.E.; Gerber, M.S.; Roux, K.H.; Sathe, S.K.J. Effects of processing on immunoreactivity of cashew nut (*Anacardium occidentale* L.) seed flour proteins. Journal of Agricultural and Food Chemistry. 2008;56 (19):8998–9005.

[42] Byun, M.W.; Lee, J.W.; Yook, H.S.; Jo, C.R.; Kim, H.Y. Application of gamma irradiation for inhibition of food allergy. Radiation Physics and Chemistry. 2002;63 (3–6):369–370.

[43] Sinanoglu, V.J.; Batrinou, A.; Konteles, S.; Sflomos, K. Microbial population, physicochemical quality and allergenicity of molluscs and shrimp treated with cobalto-60 gamma radiation. Journal of Food Protection. 2007;70 (4):958–966.

[44] Zhenxing, L.; Hong, L.; Limin, C.; Jamil, K. The influence of gamma irradiation on the allergenicy of shrimp (Penaeus vannamei). Journal of Food Engineering. 2007; 79:945–949.

[45] Leszczynska, J.; Lacka, A.; Szemraj, J.; Lukamowicz, J.; Zegota, H. The influence of gamma irradiation on the immunoreactivity of gliadin and wheat flour. European Food Research Technology. 2003;**217**:143–147.

[46] Josephson, E.S.; Peterson, M.S. Preservation of food by ionizing radiation. 2nd ed. Boca Raton: CRC Press; 1983. 170 p.